JANE'S
WARSAW PACT
Warships Handbook

JANE'S
WARSAW PACT Warships
Handbook

Edited by Captain John Moore RN
JANE'S

Cover illustration: *'Krivak I' class guided missile frigate* (Foto-Flite)

CONTENTS

INTRODUCTION

A very brief inspection of this book will show one fact very clearly – the great majority of warships and weapon systems in use in the Warsaw Pact navies are of Soviet design and many of those used in satellite fleets, of Soviet construction. Comparison with the companion volume *Jane's NATO Warships Handbook* will demonstrate one of the strengths of the Warsaw Pact at sea. While the Western alliance commanders have a medley of national fleets assigned to them, the fleets of the Pact are homogeneous with none of the incompatibility which confronts NATO in such areas as communications and stores.

Allied to this centralisation of command and design is the fact of continuity within the hierarchy. This applies not only to the political aims of the various Pact governments, which vary in degree rather than substance, but also to the commanders and designers who are able to work for many years within unchanging governmental assumptions. The record of Admiral of the Fleet of the Soviet Union Sergei Gorshkov is illustrative – having taken over as C-in-C in 1956 he was eventually relieved in December 1985. For the majority of this period he worked in close collaboration with the same senior designer.

The result of this unique aspect of continuity is the steady evolution which has taken place over the last 30 years in the main streams of various types. At the same time there have been continuing modifications to existing classes which have all taken advantage of improving technology. The most interesting aspect of these changes has been the fact that little overall capability has been lost and, in many cases, this has been markedly increased.

The overall plan has been maintained, although various diversions in design have resulted from new ideas. The Soviet designers have shown an innovative approach which is at variance with that of a number of NATO navies. In some cases this has been the result of home-grown ideas and, in others, reflects the considerable body of information culled from Western sources by both covert and open means. The difference between Western and Soviet development has, for many years, been the amount of effort, both in money and men, which the Central Committee has been prepared to divert to research and development. One of the failures of NATO authorities has long been the 'mirror image' approach – 'What we cannot achieve is clearly beyond the Soviet's capability.' This resulted in the long-held view that the Soviet Navy was ten years behind Western Fleets, a fact which was probably correct 15 years ago but is certainly not true today.

The development of the very varied types of ships for the Soviet Navy point to a determination by the command to use every facet of modern technology. While this approach has produced evolutionary designs derived from original plans it has also allowed for the simultaneous development of totally different classes.

This is best shown by the various changes that have occurred in the submarine building programme over the last 40 years. Stalin's original programme in the immediate post-war years was to build 1200 diesel submarines by 1962. Barely was this programme under way when the development of nuclear propulsion equipment began in the early 1950s. In 1958 the first Soviet nuclear submarine entered service and the construction of diesel boats was much curtailed. The 'November' class SSNs continued to join the fleet until, by 1963, 14 had been built. By this time the nuclear submarine design effort had been split – one line produced the 1967 group with pressurised water reactors; the 'Yankee' class ballistic missile submarines, the 'Charlie' class armed with cruise missiles and the 'Victor' class torpedo attack boats. During the next 20 years these original designs have evolved; the 'Yankees' into the various classes of the 'Delta' group, the 'Charlies' have become 'Charlie II' and the 'Victors' have been enlarged to become the much quieter and more capable 'Victor III' class.

A second line produced the 'Alfa' class with a titanium hull and liquid metal cooled reactors. Problems beset this design at the outset but, eventually, six boats were completed, probably test beds for new designs. The same was probably true of the fourth line, the single 'Papa' class, an improved cruise missile boat.

In the early 1980s the next generation of nuclear submarines appeared, led by the unique fifth line, the mighty 'Typhoon' armed with ballistic missiles. 'Oscar' is a direct descendant of 'Papa' but of greater size with many more missiles; 'Sierra' appears to be an improvement on 'Victor III'; 'Akula' appears to be a larger derivation of 'Alfa'; 'Mike' could well be the sixth line, a totally new concept. Thus, in 20 years, the Soviet Navy has introduced 17 different classes of nuclear submarines, some evolutionary, some revolutionary, as well as two nuclear research submarines and five classes of non-nuclear boats.

In the same 20 years the surface fleet has received notable reinforcements. The long argument about the merits of air-capable ships was resolved with the completion of the two 'Moskva' class helicopter carriers in 1967-68. In 1975 came the first of the four 'Kiev' class helicopter/VTOL carriers and in 1983 the first fixed-wing carrier, probably nuclear propelled, was laid down at Nikolayev. Backing for the carriers is provided by the nuclear-propelled battle-cruisers of the 'Kirov' class, the first of which commissioned in 1980, and the new 12 500-ton cruisers of the 'Slava' class of which three have been completed since 1982. The new generation of major surface ships is completed by the 'Sovremenny' and 'Udaloy' classes of destroyers. Each of about 8000 tons, seven of the first and nine of the latter have commissioned since 1980.

Thus a considerable fleet of brand new surface ships has appeared over the last seven years and these are backed by continuing programmes of amphibious ships and craft, missile craft, hovercraft, support ships and supply vessels as well as survey and research ships. Individual designs from East German, Polish and Romanian yards provide a measure of diversity of form but all benefit from the provision of Soviet equipment. A study of the following pages will show the Warsaw Pact navies as a well-knit and balanced fleet. This is strongly supported by the various merchant navies and fishing fleets, all operating under centralised control.

Finally, however, the reader has to remember one vitally important point. Despite technological advances the man still remains 'the greatest single factor' and all these naval ships relay on conscription to provide their junior petty officers and seamen while none of those manning the fleet has ever taken part in a naval action.

Capt J. E. Moore, RN,
September 1986

RECOGNITION SILHOUETTES

The following silhouettes are not to scale but are arranged in the descending order of classes.
No submarines and smaller surface units are covered.

'KIEV' CLASS   (USSR)

'KIROV' CLASS   (USSR)

'MOSKVA' CLASS   (USSR)

'SLAVA' CLASS   (USSR)

'KARA' CLASS   (USSR)

'KRESTA I' CLASS   (USSR)

'KYNDA' CLASS   (USSR)

'KRESTA II' CLASS   (USSR)

'SVERDLOV' CLASS   (USSR)

ADMIRAL SENYAVIN ('SVERDLOV' CLASS)   (USSR)

'SOVREMENNY' CLASS   (USSR)

'UDALOY' CLASS   (USSR)

'KASHIN' CLASS   (USSR)

'MOD.KASHIN' CLASS   (USSR)

'KILDIN' CLASS   (USSR)

'MOD.KILDIN' CLASS   (USSR)

16

'KANIN' CLASS (USSR)

'MOD.KOTLIN' CLASS (USSR)

'SAM KOTLIN' CLASS (USSR)

'SKORY' CLASS (USSR)

'KRIVAK I' CLASS (USSR)

'KONI' CLASS (USSR, GERMANY-DR)

'RIGA' CLASS (USSR, BULGARIA)

'MIRKA II' CLASS (USSR)

'GRISHA I' CLASS (USSR)

'GRISHA III' CLASS (USSR)

'PETYA I' CLASS (USSR)

'PETYA II' CLASS (USSR)

GLOSSARY

AA	Anti-aircraft
A/S, ASW	Anti-submarine warfare
CC	Cruiser
CG	Cruiser, guided missile (including surface to air missiles)
CL	Cruiser, light
CODAG, CODOG, COGAG, COGOG, COSAG	Abbreviations of mixed propulsion systems: combined diesel and gas turbine; diesel or gas turbine; gas turbine and gas turbine; gas turbine or gas turbine; steam and gas turbine
DC	Depth charge
DCT	Depth charge thrower
DD	Destroyer
DDG	Destroyer, guided missile (including surface to air missiles)
DP	Dual purpose (gun) for surface or AA use
Displacement	Basically the weight of water displaced by a ship's hull when floating: a) Light: without fuel, water or ammunition b) Standard: fully manned and stored but without fuel or reserve feed water c) Full load: fully laden with all stores, ammunition, fuel and water
FF	Frigate
FFG	Frigate, guided missile (including surface to air missiles)
Horsepower	Power developed or applied: a) bhp: brake horse power = power available at crankshaft b) shp: shaft horse power = power delivered to the propeller shaft c) ihp: indicated horse power = power produced by expansion of gases in the cylinders of reciprocating steam engines
Length	Expressed in various ways: a) oa: overall = length between extremities b) pp: between perpendiculars = between fore side of the stem and after side of the rudderpost c) wl: water-line = between extremities on the water-line
NBC	Nuclear, biological and chemical (warfare)
n miles	nautical miles
PW reactor	Pressurised water reactor
RBU	Anti-submarine rocket launcher (Soviet)
SAM	Surface to air missile
SS	Attack submarine
SSB	Ballistic missile submarine

SSBN	Nuclear-powered ballistic missile submarine
SSG	Guided missile submarine
SSGN	Nuclear-powered guided missile submarine
SSM	Surface to surface missile
SSN	Nuclear-powered attack submarine
VDS	Variable depth sonar which is lowered to best listening depth. Known as dunking sonar in helicopters

Reference Section

Number in class: 4 + ?

Displacement, tons: 25 000 tons dived
Dimensions, feet (metres): 557.6 × 75.4 × 37.7
  (170 × 23 × 11.5)
Missiles: 20—SS-N-20 tubes
Torpedo tubes: 6—21 in *(533 mm)* and 25.6 in
  (650 mm)
Main machinery: 2 nuclear reactors; 2 steam
  turbines; 2 shafts; 80 000 shp (see note)
Speed, knots: 25 approx
Complement: 150

This is the largest class of submarine ever built. The first was begun in 1977 and launched at Severodvinsk in September 1980 (in service 1983) and the second in September 1982 entering service in late 1983. The third was laid down in December 1983 for commissioning in late 1985 and the fourth followed less than a year later. The unique features of 'Typhoon' are her enormous size and the fact that the missile tubes are mounted forward of the fin. The positioning of the launch tubes forward of the fin would provide a fully integrated weapons area in the bow section leaving space abaft the fin for the provision of two nuclear reactors in each hull—probably needed to achieve a reasonable speed with this huge hull.

This submarine could operate in any ocean of the world and still have her main targets within range. In addition the configuration of the fin suggests that these could also operate through ice-cover.

'Typhoon' class (1984)

Number in class: 2 + ?

Displacement, tons: 13 600 dived
Dimensions, feet (metres): 525 × 39.4 × 28.9
 (160 × 12 × 8.8)
Missiles: 16—SS-N-23 tubes
Torpedo tubes: 6—21 in *(533 mm)* and 25.6 in
 (650 mm)
Main machinery: 2 PW nuclear reactors; 2 steam
 turbines; 60 000 shp; 2 shafts
Speed, knots: 23-24 dived
Complement: 130

First of class launched February 1984.
This class is designed for the new SS-N-23, a liquid-fuel propelled missile which carries seven MIRVs (Multiple Independent Re-entry Vehicles) to a range of 4500 n miles and is of greater accuracy than SS-N-18. They may be retrofitted in 'Delta III'. 'Delta IV' has been lengthened by 5.3 m compared to 'Delta III' and has a displacement 600 tons greater.

'Delta IV' class

Number in class: 14

Displacement, tons: 13 000 dived
Dimensions, feet (metres): 508.4 × 39.4 × 29.5
  (155 × 12 × 9)
Missiles: 16—SS-N-18 tubes
Torpedo tubes: 6—21 in *(533 mm)* (18
  torpedoes)
Main machinery: 2 PW nuclear reactors; 2 steam
  turbines; 35 000 shp; 2 shafts with 5-bladed
  propellers
Speed, knots: 24 dived
Complement: 130

The missile casing is somewhat higher than in
'Delta II' class to accommodate SS-N-18 missiles.
Built at Severodvinsk. Completed 1976-1978.
  SS-N-18 comes in three versions: Mod 1 was
the first Soviet SLBM to carry MRVs (Multiple Re-
entry Vehicles) and has a range of 3500 n miles;
Mod 2 has a single warhead with a range of 4300
nm, while Mod 3 has the same range as Mod 1
but carries seven MRVs in place of the three in
Mod 1.

'Delta III' class (1979)

Number in class: 4

Displacement, tons: 13 000 dived
Dimensions, feet (metres): 508.4 × 39.4 × 29.5
(155 × 12 × 9)
Missiles: 16—SS-N-8 tubes
Torpedo tubes: 6—21 in *(533 mm)* (bow) (18
torpedoes)
Main machinery: 2 PW nuclear reactors; 2 steam
turbines; 35 000 shp; 2 shafts
Speed, knots: 24 dived
Complement: 130

A larger edition of 'Delta I' with a straight run on
the after part of the missile casing. Building
yard—Severodvinsk. Completed 1975. Probably
an interim class while the final trials of SS-N-18
were underway and SS-N-23 was being
developed.

Left: *'Delta II' class (1982)*
Right: *'Delta II' class (US Navy, 1981)*

Number in class: 18

Displacement, tons: 11 000 dived
Dimensions, feet (metres): 455.9 × 39.4 × 29.5
  (139 × 12 × 9)
Missiles: 12—SS-N-8 tubes
Torpedo tubes: 6—21 in *(533 mm)* (bow) (18
  torpedoes)
Main machinery: 2 PW nuclear reactors; 2 steam
  turbines; 35 000 shp; 2 shafts
Speed, knots: 26 dived
Complement: 120

The first of this class, an advance on the 'Yankee' class SSBNs was laid down at Severodvinsk in 1969 and completed in 1972. The longer-range SS-N-8 missiles are of greater length than the SS-N-6s and, as this length cannot be accommodated below the keel, they stand several feet proud of the after-casing. At the same time the need to compensate for the additional top-weight would seem to be the reasons for the reduction to 12 missiles in this class. Programme completed 1972-77. Building yards— Severodvinsk (10) and Komsomolsk (8).

'Delta I' class (1976)

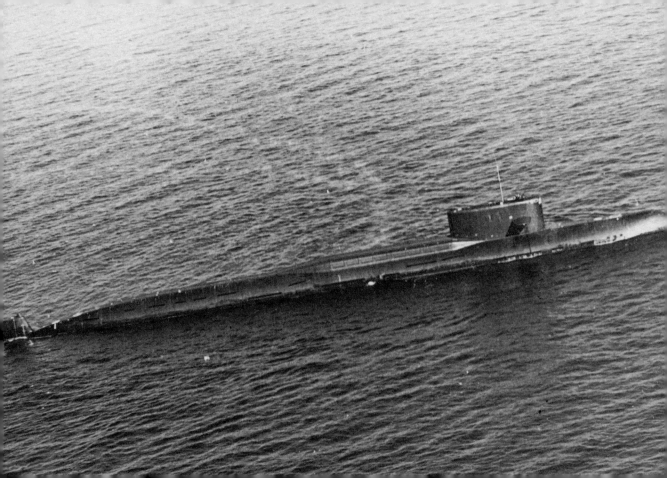

Number in class: 19 + 1 + 14

(See 'Class' note)

Displacement, tons: 10 500 dived
Dimensions, feet (metres): 426.4 × 38 × 26.2
(132 × 11.6 × 8)
Missiles: 16—SS-N-6 tubes (see note)
Torpedo tubes: 6—21 in *(533 mm)*; 18 torpedoes
carried
Main machinery: 2 nuclear reactors; 2 steam
turbines; 35 000 shp; 2 shafts
Speed, knots: 28 dived
Complement: 120

The vertical launching tubes are arranged in two rows of eight, and the SS-N-6 missiles have a range of 1 300/1 600 n miles. The Mod 3 version of these missiles have two MRV warheads as in the first of the SS-N-18 Mod 1. At about the time that the decision to convert USS *Scorpion* into the world's first SSBN hull, to be named *George Washington*, was taken on 31 December 1957 it is likely that the Soviet Navy embarked on its own major SSBN programme. With experience gained from the diesel-propelled 'Golf' class and the nuclear-propelled 'Hotel' class, the 'Yankee' design was completed, modelled on the USS 'Ethan Allen' design. The first of the class was laid down in 1963-64 and delivered late 1967 and the programme then accelerated with output rising to six to eight a year in the period around 1970. The last boat was completed in 1974. Construction took place at Severodvinsk (first laid down in 1965) and Komsomolsk. The original

deployment of this class was to the Eastern seaboard of the US giving a coverage at least as far as the Mississippi. Increase in numbers allowed a Pacific patrol to be established off California providing coverage from the western seaboard to the eastern side of the Rockies.

Appearance: Fin mounted fore-planes.

Class: Of the total of 34 built, 19 are fitted with SS- N-6 Mod 1 (1 300 n miles), or with SS-N-6 Mod 3 (1 600 n miles) with 12 SS-N-17 (2 100 n miles—solid propellant with one warhead) ('Yankee II'). The others have had their missile tubes removed, the first in 1978 and will be converted for other tasks, including SSN. One has been converted for the long range land attack cruise missile SS-NX-24.

'Yankee I' class (1978)

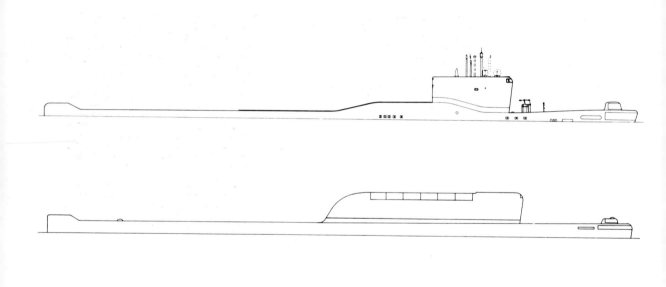

Number in class: 1

Displacement, tons: 6 500 dived
Dimensions, feet (metres): 426.4 × 29.8 × 25
  (130 × 9.1 × 7.6)
Missiles: 6—SS-N-8 tubes
Torpedo tubes: 6—21 in *(533 mm)* (bow); 2—
  16 in *(406 mm)* (stern); 20 torpedoes carried
Main machinery: 2 nuclear reactors; 2 steam
  turbines; 30 000 shp; 2 shafts
Speed, knots: 22 dived
Complement: 90

All this class was completed as 'Hotel I' class at Severodvinsk between 1958 and 1962. Originally fitted with SS-N-4 system with Sark missiles (300 n miles) with three vertical ballistic missile tubes in the large fin. Between 1963 and 1970 this system was replaced at the building yard by the dived launch SS-N-5 system with single-headed Serb missiles capable of 750 n mile range, the class then being renamed 'Hotel II'. A number of these boats was deployed off both East and West coasts of the USA and Canada. The 'Hotel III' is a single unit converted in 1969-70 at Severodvinsk for the test firings of the SS-N-8 (carries six missiles with an extended hull and fin). This class was of a similar hull and reactor design to the 'Echo' class.

Class: 'Hotel II' have had their missile tubes removed and of the remaining seven one was converted to a 'communications submarine' in the late 1970s (SSQN) and the remainder have been or are being converted to SSN.

Left above: *'Hotel II' SSQN
(communications) 1985*
Left below: *'Hotel III' 1985*

Number in class: 13 + 1 + 1

Displacement, tons: 3 000 dived ('Golf II' and 'V'); 3 500 dived ('Golf III')
Dimensions, feet (metres): 324.8 × 26.2 × 24.6 *(99 × 8 × 7.5)* ('Golf III', 390.4 *(119)*)
Missiles: 3—SS-N-5 tubes ('Golf II'); 6—SS-N-8 tubes ('Golf III'); 1—SS-N-20 tube ('Golf V')
Torpedo tubes: 6—21 in *(533 mm)* (4 bow, 2 stern); 12 torpedoes carried
Main machinery: Direct drive or diesel-electric; 3 diesels; 6 000 hp; electric motors; 5 500 hp; 3 shafts
Speed, knots: 15 surfaced; 14 dived
Range, miles: 7 000 surfaced max
Complement: 87 (12 officers, 75 men)

The 'Golf II' class has a very large fin fitted with three vertically mounted tubes and hatches for launching ballistic missiles. Twenty-three 'Golf I' class were built in 1958-62 at Komsomolsk (8) and Severodvinsk (15). Of these 13 were converted to 'Golf II' class (8 at Severodvinsk 1963-68 and 5 at Komsomolsk 1963-68) to carry the SS-N-5 system. This system, with a range of 750 n miles using single-headed Serb missiles, replaced the SS-N-4 system with 300 n miles Sark missiles.

One sank in the Pacific in 1968 and was partially raised by the USA in 1974. Three scrapped in late 1970s.

One of this class has been built by China.

Class: Considerable modifications have been made to boats of the 'Golf I' class. Those undergoing major changes have been redesignated as follows:

'Modified Golf I': Three have had missiles removed and now serve as what are probably communications submarines.

'Golf III': The hull and fin have both been extended by 65.6 ft *(20 m)* to accommodate six SS-N-8 launch tubes in the fin.

'Golf V': Converted for research and development on the SS-N-20 in 1974-75 at Severodvinsk.

'Golf II' class (1979)

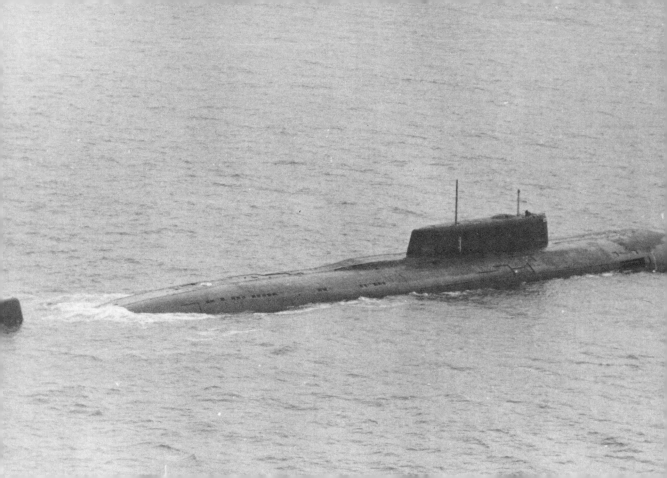

Number in class: 3 + 1

Displacement, tons: 14 000 dived
Dimensions, feet (metres): 492 × 60 × 36 *(150 × 18.3 × 11)*
Missiles: 24 tubes for SS-N-19
Torpedo tubes: 6—21 in *(533 mm)* and 25.6 in *(650 mm)* (18 reloads)
Speed, knots: 33 dived
Complement: 130

First laid down at Severodvinsk in 1978 and launched in spring 1980. Started trials late 1980. Second launched 1982 and third in 1985. The missile tubes are carried in the casing, resulting in an unusually large beam. Two reactors could provide the power for about 30 knots. With this speed and a formidable armament this class could well operate on the bow of a task group. (see 'Papa' class notes, p. 41)

'Oscar' class (1985)

Number in class: 1

Displacement, tons (approx): 8 000 dived
Dimensions, feet (metres): 357.5 × 37.7 × 24.9
 (109 × 11.5 × 7.6)
Missiles: 10 tubes for SS-N-9
Torpedo tubes: 4—21 in *(533 mm)* (bow)
Main machinery: 2 nuclear reactors; 2 steam
 turbines; 60 000 shp; 2 shafts
Speed, knots: 37 dived
Complement: 90

A single member of an SSGN class, apparently a development of the 'Charlie' class. Built 1969-71. The fin is of a much more angular shape than the 'Charlie' classes with a higher casing, a more rounded bow and with the missile tubes having square covers. Built at Severodvinsk.

The hull is of titanium alloy, the design probably benefitting from experience gained with the 'Alfa' class (p. 58). The introduction of the SS-N-9 missile with a range of 60 n miles was the forerunner of the new equipment to be fitted in 'Charlie II'. 'Papa' was apparently a successful design, that of 'Oscar' being very similar, though larger.

'Papa' class (1975)

Number in class: 6

Displacement, tons: 5 500 dived
Dimensions, feet (metres): 334.6 × 32.5 × 25.6
  (102 × 9.9 × 7.8)
Missiles: 8 tubes for SS-N-9
Torpedo tubes: 6—21 in *(533 mm)* (14 torpedoes
  carried)
Main machinery: 1 PW nuclear reactor; 1 steam
  turbine; 15 000 shp; 1 shaft
Speed, knots: 24 dived
Complement: 90

Enlarged 'Charlie' class. Built at Gorky from
1973-80.
  The increase in size may be due to the fitting of
improved command and control equipment for
the submerged launched SS-N-9, first seen in
'Papa' class.

Left: *'Charlie II' class (1980)*
Right: *'Charlie II' class*

Number in class: 11

Displacement, tons: 4 800 dived
Dimensions, feet (metres): 313.2 × 32.5 × 24.6
  (95.5 × 9.9 × 7.5)
Missiles: 8 tubes for SS-N-7
Torpedo tubes: 6—21 in *(533 mm)* (bow); 14
  torpedoes carried
Main machinery: 1 PW nuclear reactor; 1 steam
  turbine; 15 000 shp; 1 shaft
Speed, knots: 24 dived
Complement: 90

A class of cruise missile submarines built at Gorky 1967-72. With eight missile tubes for the SS-N-7 system (35 n mile range) which has a dived launch capability, they represented a significant achievement in the submarine cruise missile field. These boats have an improved hull and reactor design and must be assumed to have an organic control for their missile system therefore posing a notable threat to any surface force. Their deployment to the Mediterranean, the area of the US 6th Fleet, is only a part of their general and world-wide operations.

One sank off Petropavlovsk in June 1983 and was subsequently salved.

Appearance: Although similar to the 'Victor' class the bulge at the bow, the almost vertical drop of the forward end of the fin, a slightly lower after casing and a different arrangement of free-flood holes in the casing give a clear difference.

'Charlie I' class (1978)

Number in class: 29

Displacement, tons: 6 200 dived
Dimensions, feet (metres): 380.5 × 32.1 × 23.9
  (116 × 9.8 × 7.3)
Missiles: 8 tubes for SS-N-12/3A
Torpedo tubes: 6—21 in *(533 mm)* (bow); 2—
  16 in *(406 mm)* (stern); 20 torpedoes carried
Main machinery: 2 nuclear reactors; 2 steam
  turbines; 30 000 shp; 2 shafts
Speed, knots: 24 dived
Complement: 100

The decision to produce this class may have been due to the availability of building ways because of a break in SSBN production between the 'Hotel' and 'Yankee' classes in the first half of the 1960s, the development of the SS-N-3A (220 n miles) cruise missile with an anti-ship capability and the need to counter the threat from Western strike carriers.

With a slightly lengthened hull over the 'Echo I', a fourth pair of launchers was installed and between 1961-67 29 of this class were built. They are now split evenly between the Pacific and Northern fleets and still provide a useful capability, particularly those modified to carry the SS-N-12 missile. This class is often deployed to the Mediterranean and Indian Ocean.

Built at Severodvinsk and Komsomolsk.

'Echo II' class (1979)

Number in class: 16

Displacement, tons: 3 500 surfaced; 4 300 dived
Dimensions, feet (metres): 284.4 × 33.1 × 22.9
(86.7 × 10.1 × 7)
Missiles: 4 tubes for SS-N-3A; 2 before and 2
abaft the fin
Torpedo tubes: 6—21 in *(533 mm)* (bow); 18
torpedoes carried
Main machinery: Direct drive or diesel-electric; 3
diesels; 3 500 bhp; 2 electric motors; 3 500 hp;
2 shafts
Speed, knots: 12 surfaced; 12 dived
Complement: 79

Completed between 1961 and 1968 at Gorky.
Four SS-N-3A launchers, one pair either end of
the fin which appears to be comparatively low.
This class was the logical continuation of the
'Whiskey' class conversions. This class is
deployed to the Mediterranean and has been
deployed to the Indian Ocean. In 1980- 81 four of
this class were transferred from the Northern
Fleet to the Baltic Fleet, probably establishing a
pattern of patrols in that area.

Appearance: The massive casing and the long,
fairly low fin make this an unmistakable class.

*'Juliett' class (Royal Danish Navy,
1982)*

Number in class: 1

Displacement, tons: 1 200 surfaced; 1 500 dived
Dimensions, feet (metres): 274.9 × 21.3 × 16.4
 (83.8 × 6.5 × 5)
Missiles: 4 tubes for SS-N-3C
Torpedo tubes: 4—21 in *(533 mm)* (bow); 8
 torpedoes carried
Main machinery: Direct drive or diesel-electric;
 2 diesels; 4 000 bhp; electric motors; 2 700 hp;
 2 shafts
Speed, knots: 17 surfaced; 12 dived
Range, miles: 9 000 surfaced at 10 knots
Complement: 60

A more efficient modification of the 'Whiskey'
class than the earlier 'Twin-Cylinder' with four
SS-N-3C launchers built into a remodelled fin on
a hull lengthened by 25 ft. Converted between
1960-63. Has little or no capability against ship
targets. Must be a very noisy boat when dived.
Last boat based in the Baltic.
 The last 'Twin-Cylinder' was probably scrapped
in 1983.

*'Whiskey Long-Bin' class (US Navy,
1973)*

Number in class: 1 + 1

Displacement, tons: 8 000 dived
Dimensions, feet (metres): 351 × 36.7 × 24.6
 (107 × 11.2 × 7.5)
Torpedo tubes: 6—21 in *(533 mm)* and 25.6 in
 (650 mm) for torpedoes, SS-NX-21, SS-N-16
 and 15
Main machinery: Nuclear
Speed, knots: 30+ dived

First of class launched mid-1984. This class is probably the successor to the 'Victor III'. The very long fin is particularly notable.

The first of class was built at Komsomolsk and some reports suggest that the hull may be of titanium alloy. There is a remarkable similarity between 'Akula' and 'Alfa' (p. 58), the tormer being approximately 1.3 times the size of the latter while the proportion of the fin length to the overall length is the same (1:4). The large pod on the tail fin is a continuation of the 'Victor III' design (p. 60).

'Akula' class (DoD, 1985)

Number in class: 1 + 1

Displacement, tons: 9 700 dived
Dimensions, feet (metres): approx 360.8 × 39.4
  × 29.5 *(110 × 12 × 9)*
Torpedo tubes: 6—21 in *(533 mm)* and 25.6 in
  (650 mm) for torpedoes and SS-NX-21 (see
  note)
Main machinery: 2 nuclear reactors (? liquid
  metal cooled); 1 steam turbine; 50-60 000 shp;
  1 shaft
Speed, knots: 35+ dived
Complement: approx 95

First launched at Severodvinsk May 1983, in
service late 1984. Possibly titanium hulled and
the result of experience with the 'Alfa' class. The
length to beam ratio of the hull (10 to 1) is,
however, very different from 'Alfa'. There is no
pod on the tail fin and this design may be a
separate development.

Missiles: This class can probably carry SS-N-15,
SS-N-16 and 1 600 n mile range SS-NX-21
missiles in place of torpedoes.

*'Mike' class (Royal Norwegian Air
Force, 1985)*

Number in class: 1 + 1

Displacement, tons: 8 000 dived
Dimensions, feet (metres): 352.6 × 40.7 × 24.3
  (107.5 × 12.4 × 7.4)
Torpedo tubes: 6—21 in *(533 mm)* and 25.6 in
  (650 mm) for torpedoes and SS-NX-21
Main machinery: 2 PW nuclear reactors; 1 steam
  turbine; approx 40 000 hp; 1 shaft
Speed, knots: 32 approx dived
Complement: approx 100

First launched July 1983 at Gorky, in service late 1984. This may be the true successor to 'Victor III' and this class may be in rapid series production in the near future. Use may be made of at least one more building yard, possibly Admiralty Yard, Leningrad.

Diving depth: ? 1 800 ft *(550 m)* operational.

'Sierra' class (1984)

Number in class: 6

Displacement, tons: 3 700 dived
Dimensions, feet (metres): 259.1 × 32.8 × 24.9
(79 × 10 × 7.6)
Torpedo tubes: 6—21 in *(533 mm)*
Main machinery: 2 liquid metal cooled nuclear
reactors; 2 steam turbo-alternators; 1 motor;
turbo-electric; 47 000 shp; auxiliary diesel; 1
shaft
Speed, knots: 42+ dived
Complement: 40

The first of this class was laid down in mid-1960s
and completed in 1970 at Sudomekh, Leningrad.
The building time was very long in comparison
with normal programmes and it seems most
likely that this was a prototype. This boat was
scrapped in 1974. The reduction of the length
combined with the high speed shows that a new
type of reactor has been developed, probably
liquid metal cooled, and this power probably
requires the use of a super-conducting motor. A
greater diving depth, possibly down to 2 500 ft
(700 m) has been achieved by use of titanium
alloy for the hull. Production was slow with six
more completed by 1983 at Sudomekh and
Severodvinsk, the programme then being
completed. All these six boats may have minor
variations, possibly to study future designs now
at sea in the 'Mike' class.

'Alfa' class (1984)

Number in class: 21

Displacement, tons: 6 300 dived
Dimensions, feet (metres): 341.1 × 32.8 × 23
  (104 × 10 × 7)
Torpedoes: 6—21 in *(533 mm)* and 25.6 in
  (650 mm); 18 torpedoes carried (see note)
Main machinery: 2 PW nuclear reactors; 1 steam
  turbine; 30 000 shp; 1 shaft; 2 auxiliary
  propellers
Speed, knots: 30 dived
Complement: 100

An improvement on 'Victor II', the first of class being completed at Komsomolsk in 1978. Since then building has also been carried out at Admiralty Yard, Leningrad and this has been a very rapid building programme which was completed with the launch of the last boat in 1985. The first of this class was very different from her predecessors in having a large pod on the tail-fin. The purpose of this pod (repeated in 'Sierra' and 'Akula') has been a matter of discussion. The first two suggestions, that this is either for a towed array or for the launching of decoys, ignore the fact that this is a well streamlined submarine and the addition of a protruberance which would cause extra turbulence is unlikely if the tasks could be executed without this large lump being slung high above the stern. If however, this is an auxiliary, slow-speed and silent propulsor on the magneto-hydrodynamic thrust principle (MHD) the disadvantages might be considered worth while.

Diving depth: 1 300 ft *(400 m* approx*)* operational; 2 000 ft *(600 m* approx*)* crushing

'Victor III' class in 1983 (via John Berg)

Number in class: 7

Displacement, tons: 6 000 dived
Dimensions, feet (metres): 334.6 × 32.8 × 22.3
   (102 × 10 × 6.8)
Torpedo tubes: 6—21 in *(533 mm)* (bow) and
   25.6 in *(650 mm)*; 18 torpedoes carried
Main machinery: 2 PW nuclear reactors; 1 steam
   turbine; 30 000 shp; 1 shaft; 2 auxiliary
   propellers
Speed, knots: 29 dived
Complement: 100

An enlarged 'Victor I' design, 9 metres longer.
This additional length may be used to house the
equipment needed for firing the tube-launched
SS-N-15, a Subroc type weapon. First appeared
in 1972—class completed 1978. Built at
Admiralty Yard, Leningrad and Gorky.

Diving depth: 1 300 ft *(400 m* approx*)*
operational.

'Victor II' class

Overleaf
Left: *'Victor' class (1974)*
Right: *'Victor' class (1975)*

Number in class: 16

Displacement, tons: 5 300 dived
Dimensions, feet (metres): 305.1 × 32.8 × 23 *(93 × 10 × 7)*
Torpedo tubes: 6—21 in *(533 mm)* (bow); 18 torpedoes carried
Main machinery: 2 PW nuclear reactors; 1 steam turbine; 30 000 shp; 1 shaft; 2 auxiliary propellers
Speed, knots: 29 dived
Complement: 90

Designed purely as an attack submarine for both A/S and anti-ship roles. The first of class laid down in 1965 entering service in 1967-68—class completed 1974 at a building rate of two per year. Superseded by the 'Victor II' programme. Built at Admiralty Yard, Leningrad.

This was the first Soviet submarine with an Albacore hull-form and a new reactor system, a new generation design shared by the 'Charlie' class.

The majority is deployed with the Northern Fleet, although some have joined the Pacific Fleet.

Diving depth: 1 300 ft *(400 m* approx*)* operational.

Left: *'Victor I' class*

Number in class: 5

Displacement, tons: 5 500 dived
Dimensions, feet (metres): 367.4 × 32.2 × 25.6
(112 × 9.8 × 8)
Torpedo tubes: 6—21 in *(533 mm)* (bow); 2—16
in *(406 mm)* (stern); 20 torpedoes carried
Mines: Can lay 36 mines in place of torpedoes
Main machinery: 2 nuclear reactors; 2 steam
turbines; 30 000 shp; 2 shafts
Speed, knots: 28 dived
Complement: 92 (12 officers, 80 men)

This class was completed in 1960-62 at
Komsomolsk. Originally mounted six SS-N-3
launchers in the after casing.

The hull of this class is very similar to the
'Hotel'/'November' type and it is probably
powered by similar nuclear plant. Only five 'Echo I'
class were built, being followed immediately by
the 'Echo II' class. In 1969-74 the 'Echo I' class
was converted into fleet submarines with the
removal of the missile system, a decision made
possible by the development of the 'Yankee' class
with its strategic missile system.

'Echo I' class

Number in class: 12

Displacement, tons: 4 200 surfaced; 5 000 dived
Dimensions, feet (metres): 359.8 × 29.8 × 21.9 *(109.7 × 9.1 × 6.7)*
Torpedo tubes: 8—21 in *(533 mm)* (bow); 2—16 in *(406 mm)* (stern); 26 torpedoes carried
Main machinery: 2 nuclear reactors; 2 steam turbines; 30 000 shp; 2 shafts
Speed, knots: 30 dived
Complement: 86

The first class of Soviet nuclear submarines which entered service between 1958 and 1963. Built at Severodvinsk. The hull form with the great number of free-flood holes in the casing suggests a noisy boat. In April 1970 one of this class sank south-west of the UK and another was probably scrapped in the late 1970s.

Diving depth: Reported as 1 150 ft *(350 m)*.

'November' class (MoD, 1979)

Number in class: 2 + 4

These six boats had their missile capability removed in the early 1980s so that the SLBM total remained within the SALT I limit of 950. Two completed conversion to SSN, remainder in hand.

Left: *'Hotel II' class (1982)*
Right: *'Hotel II' class*

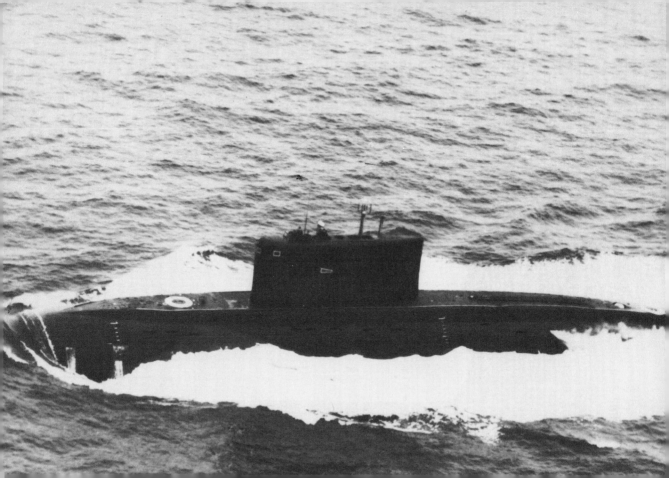

Number in class: 10 + 3

Displacement, tons: 2 500 surfaced; 3 200 dived
Dimensions, feet (metres): 229.6 × 29.5 × 23 *(70 × 9 × 7)*
Torpedo tubes: 8—21 in *(533 mm)*
Main machinery: Diesel-electric
Speed, knots: 16 dived
Complement: ? 55

First probably launched in 1979 at Komsomolsk. Construction is now taking place in Gorky and Leningrad as well as at Komsomolsk. At Leningrad the 'Kilo' programme is replacing the 'Foxtrot' export stream and the first was delivered to India in early 1986. India and other client states may follow.

The building rate for the Soviet Navy has accelerated markedly and is now equal to the 'Foxtrot' programme at its peak. There is evidence of trials being carried out with a SAM launcher fitted on the fin.

'Kilo' class (1982)

Number in class: 18

Displacement, tons: 3 000 surfaced; 3 900 dived
Dimensions, feet (metres): 301.8 × 29.5 × 23 *(92 × 9 × 7)*
Torpedo tubes: 8—21 in *(533 mm)*
Main machinery: Diesel-electric; 3 diesel-generators; 6 000 shp; 2 electric motors; 5 500 shp; 2 shafts
Speed, knots: 15 surfaced; 16 dived
Complement: 62

This class was first seen at the Sevastopol Review in July 1973 and, immediately succeeding the 'Foxtrot' class in 1972, showed a continuing commitment to non-nuclear-propelled boats. An improved design intended for long-range operations as shown by current deployment in the Mediterranean. The building rate rose to two a year at Gorky and the programme is now finished. Stationed only in the Northern and Black Sea Fleets.

'Tango' class (1976)

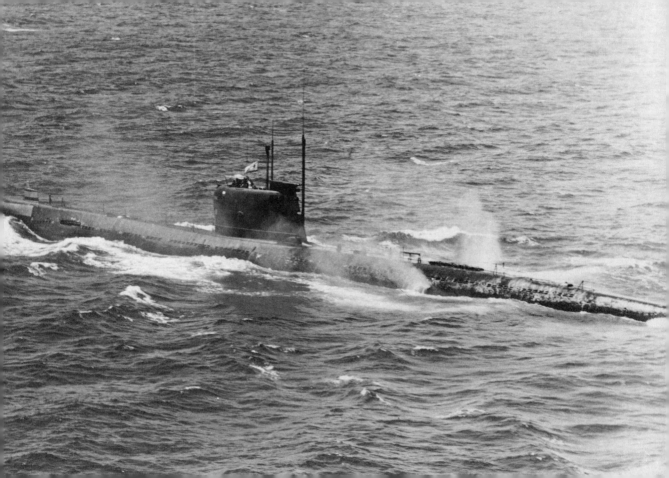

Number in class: 60

Displacement, tons: 1 950 surfaced; 2 500 dived
Dimensions, feet (metres): 300.1 × 26.2 × 20
  (91.5 × 8 × 6.1)
Torpedo tubes: 6—21 in *(533 mm)* (bow); 4—16
  in *(406 mm)* (stern); 22 torpedoes carried
Main machinery: Diesel-electric; 3 diesels; 6 000
  bhp; 3 electric motors; 5 500 hp; 3 shafts
Speed, knots: 16 surfaced; 16 dived
Range: 20 000 miles surfaced, cruising
Complement: 75

Built between 1958 and 1971 at Sudomekh for
the Soviet Navy. Production continued until 1984
for transfer to other countries eg Cuba, India,
Libya. A follow-on of the 'Zulu' class. Only 62 out
of a total programme of 160 were completed as
the change-over to nuclear boats took effect. A
most successful class which has been deployed
world-wide, forming the bulk of the Soviet
submarine force in the Mediterranean in the
1960s and 1970s. Two boats believed to have
been scrapped following accidents.

'Foxtrot' class (1971)

Number in class: 2 (+ 4 Reserve)

Displacement, tons: 1 950 surfaced; 2 300 dived
Dimensions, feet (metres): 295.2 × 24.3 × 20 *(90 × 7.4 × 6.1)*
Torpedo tubes: 10—21 in *(533 mm)* (6 bow, 4 stern); 22 torpedoes carried or 36 mines and 4 torpedoes
Main machinery: 3 diesels; 6 000 bhp; 3 electric motors; 5 500 hp; 3 shafts
Speed, knots: 16 surfaced; 16 dived
Range, miles: 20 000 surfaced, cruising
Complement: 75

The first large post-war patrol submarines built by USSR. Twenty-six completed from late 1951 to 1955 out of an original programme of 40. General appearance is streamlined with a complete row of free-flood holes along the casing. Some were fitted with High Test Peroxide (HTP) machinery for high underwater speeds up to 17 knots, but this was removed as the inflammatory nature of the fuel was appreciated. Eighteen were built by Sudomekh Shipyard, Leningrad, in 1952-55 and others at Severodvinsk. The general external similarity to the later German U-boats of the Second World War is notable. All now appear to be of the 'Zulu IV' type. This class is obsolescent and numbers are steadily declining.

The six 'Zulu V' conversions of this class provided the first Soviet ballistic missile submarines with SS-N-4 systems, the first being completed in 1955. These conversions no longer have a missile capability and have either been scrapped or are used in an auxiliary role such as oceanographic research.

'Zulu IV' class (1981)

Number in class: 6

Displacement, tons: 1 400 surfaced; 1 700 dived
Dimensions, feet (metres): 251.9 × 23.9 × 18
(76.8 × 7.3 × 5.5)
Torpedo tubes: 8—21 in *(533 mm)* (6 bow, 2
  stern); 14 torpedoes or 28 mines in place of
  torpedoes
Main machinery: 2 diesels; 4 000 bhp; 2 electric
  motors; 4 000 hp; 2 shafts
Speed, knots: 16 surfaced; 13 dived
Range, miles: 19 000 at 9 knots (surfaced)
Complement: 54

These are an improved 'Whiskey' class design
with modernised conning tower, and sonar
installation. All built in 1958-61 at Gorky. This
was to have been numerically the largest class in
the post-war submarine build-up. As their
construction period coincided with the successful
introduction of nuclear propulsion only about 20
were completed out of the staggering planned
total of 560. Some of this class are used for
experimental work.

Transfers: Six to Egypt in 1966-68, two to
Bulgaria in early 1970s and one in 1984 , one to
Algeria in January 1982 and one in February
1983 and two to Syria in 1985. China and North
Korea building submarines of similar design.

'Romeo' class (1980)

Number in class: 50 (+ 65 Reserve)

Displacement, tons: 1 080 surfaced; 1 350 dived
Dimensions, feet (metres): 249.3 × 21.3 × 16.1
  (76 × 6.5 × 4.9)
Torpedo tubes: 6—21 in *(533 mm)* (4 bow, 2
  stern); 12 torpedoes carried (or 24 mines)
Main machinery: Direct drive or diesel-electric; 2
  diesels; 4 000 bhp; 2 electric motors; 2 700 hp;
  2 shafts
Speed, knots: 18 surfaced; 14 dived
Range, miles: 8 500 at 10 knots (surfaced)
Complement: 54

This was the first post-war Soviet design for a
medium- range submarine. Like its larger
contemporary the 'Zulu', this class shows
considerable German influence. About 240 of the
'Whiskey' class were built between 1951 and
1957 at Gorky, Komsomolsk, Leningrad (Baltic
Yard) and Nikolayev. Like its successor, the
'Romeo', this class was cut back from the original
planned total of 340 as nuclear propulsion
became established. Built in five types—I and IV
had guns forward of the conning tower, II had
guns both ends, while III and V have no guns. All
Soviet 'Whiskey' class are now type V.

Class total: There may be about 50 operational
although this class is rarely seen out of area.

Conversions: Two of this class, named
Severyanka and *Slavyanka,* were converted for
oceanographic and fishery research and have
probably now been scrapped.

Foreign transfers: Has been the most popular
export model, with 40 in all transferred. Currently
in service: Albania (3), Cuba (1) (non operational),
Egypt (4), Indonesia (2), North Korea (4),
Poland (3). China has carried out her own
building programme.

'Whiskey V' class (1982)

Number in class: 2

Displacement, tons: 4 000 dived
Dimensions, feet (metres): 351 × 32.8 × 23 *(107 × 10 × 7)*
Main machinery: 2 diesels; ? 6 000 shp; 3 electric motors; ? 6 000 shp
Speed, knots: 15 surfaced; 10 dived

One is in service in the Pacific and one in the Northern Fleet. Designed for rescue work and carry two DSRVs (Deep Submergence Recovery Vehicles) on the after casing.

Left: *'India' class (1984)*
Right: *'India' class in the Pacific (US Navy)*

Number in class: 4

Displacement, tons: 3 000 dived
Dimensions, feet (metres): 239.5 × 32.1 × 23.9
   (73 × 9.8 × 7.3)
Torpedo tubes: 4—21 in *(533 mm)* (bow)
Main machinery: Diesel-electric; 1 diesel; 4 000
   hp and 1 diesel generator; 1 electric motor;
   4 000 hp; 1 shaft
Speed, knots: 15 dived
Complement: 60

The beam-to-length ratio is larger than normal in a diesel submarine which would account in part for the large displacement for a comparatively short hull.

First completed in 1968 at Komsomolsk. These four submarines are in the Northern, Black Sea and Pacific fleets and act as 'padded targets' for ASW exercises and weapon firings.

'Bravo' class (1984)

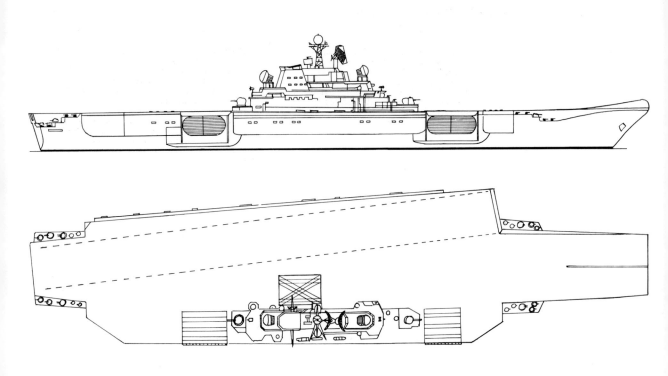

Number in class: 2

Displacement, tons: 65-75 000 (approx)
Dimensions, feet (metres): 984 × 125 wl × 36
 approx *(300 × 38 × 11)*
Aircraft: 50 (approx)
Main engines: Possibly 2 nuclear reactors; steam
 turbines; 4 shafts; 200 000 shp approx
Speed, knots: 32

The building of the first ship, a logical continuation of the 'Kiev' class and a basic component of a task force including the nuclear-propelled 'Kirov' class battle cruisers, was reported as acknowledged by Admiral Gorshkov in 1979. Nikolayev has the facilities and experience gained on the 'Kiev' class. Sea trials of first ship could take place in 1988 and her sister two to three years later but the operation of a CTOL carrier is a complex operation to be learned from scratch. The first may be operational by 1992.

Nuclear-powered new construction

Number in class: 4

BAKU
KIEV
MINSK
NOVOROSSIYSK

Novorossiysk ('Kiev' class) (MoD, 1983)

Displacement, tons: 37 100 full load
Dimensions, feet (metres): 895.7 × 107.3 (hull wl) × 32.8 (screws) *(273 × 32.7 × 10)*
Beam, feet (metres): 154.8 *(47.2)* (oa, including flight deck and sponsons)
Aircraft (estimated): 32—normal mix, 12 Forger A, 1 Forger B, 16 Hormone A or Helix, 3 Hormone B
Missiles: SSM; 4 twin SS-N-12 (16 reloads) SAM; 2 twin SA-N-3 (72 missiles); 2 twin SA-N-4 (40 missiles) (see note)
Guns: 4—76 mm (twin); 8 Gatling 30 mm mounts
A/S weapons: 1 twin SUW-N-1 (for FRAS-1); 2—12-barrelled RBU 6000 launchers fwd
Torpedo tubes: 10—21 in *(533 mm)* (quin)
Main engines: 4 steam turbines; 4 shafts (2 rudders); 200 000 shp
Speed, knots: 32
Fuel, tons: 7 000
Range, miles: 13 500 at 18 knots; 4 000 at 31 knots
Complement: 1 200 plus air group
Commissioned: 1975-86

The first post-war sign of Soviet acceptance of the need for organic air was the appearance of *Moskva* and *Leningrad* in 1968-69 (see 'Moskva' class p. 96). This class was the first to be built with a surface warfare capability as well as being ASW ships. They carried the embarked helicopter concept a long stage further than the cruisers with a single embarked helicopter. It is most probable that a much larger number of ships was projected and the reason for the cancellation of the remainder might be because of any or all of a number of factors. Two appear to be of considerable importance—the growing Soviet realisation of the important part its Navy could play in overseas affairs and the appearance of the prototype of the first Soviet VTOL aircraft, the Yakovlev Freehand. This first appeared in public in 1967 and its capabilities were known at least a year before that. This was ten years before *Kiev* became operational, *(Kiev* passed the Turkish Straits on 18 July 1976) a reasonable lead time for Soviet designers and constructors.

The task of this class is probably twofold—an advanced ASW role in wartime and an intervention capability in so-called peacetime with excellent command, control and communication capabilities. The inclusion of SS-N-12 (295 n mile) capability, A/S weapons as well as sonar equipment, and a gun armament as well as both SAM and Point Defence Systems shows the initiation of Soviet requirement for multi-purpose ships.

These are probably all that will be built to this design although a follow-on class of conventionally propelled carriers cannot be ruled out.

Number in class: 2 + 1

FRUNZE
KIROV

Displacement, tons: 22 000 standard; 28 000 full load
Dimensions, feet (metres): 813.6 × 93.5 × 29.5 *(248 × 28.5 × 9.1)*
Aircraft: 3 Hormone or Helix helicopters
Missiles: SSM; 20—SS-N-19 (no reloads); SAM; 12—SA-N-6 (96 missiles); 2—SA-N-4 (twin) (40 missiles); SAM; 16 SA-NX-9 launchers *(Frunze only—128 missiles)*; ASW; 2—SS-N-14 (twin) (14 missiles) (not in *Frunze*)
Guns: 2—100 mm (single) *(Kirov)*; 2—130 mm (twin) *(Frunze)*; 8 Gatling 30 mm
Torpedo tubes: 8—21 in *(533 mm)* (quad)
A/S weapons: 1—12-barrelled RBU 6000 (forward); 2—6-barrelled RBU 1000 (aft) (not in *Frunze*)
Main engines: 2 nuclear reactors and oil-fired superheat boilers for steam turbines; 2 shafts; 150 000 shp
Speed, knots: 33
Complement: 900
Commissioned: 1980 -

The first Soviet surface combatant with nuclear propulsion, this class, of similar dimensions and speed to the battle cruisers of the past, is a continuation of the Soviet plan for large dual-purpose ships first seen in the 'Kiev' class. The design is a self-contained element coming between the 'Kara' and 'Kiev' with a much-enhanced range due to nuclear propulsion. The similarity between this design and that of the US Navy's Strike Cruiser is apparent. The tasks of the latter, 'Screening ships for carriers in high threat areas but also undertaking independent operations', could also be true of the 'Kirov' class. As the support for a nuclear carrier or as the focus of a task force including the 'Kiev', 'Slava', 'Kara', 'Krivak' and 'Ivan Rogov' classes with support from 'Berezina' or 'Boris Chilikin' classes they would form a formidable intervention force, with VTOL and helicopter aircraft as well as a full outfit of missiles and heavy guns. *Frunze* has a modified superstructure and armament although the SS-N-19 and SA-N-6 missile fits are the same. The guns are a twin 130 mm mounting similar to the new mountings in the 'Sovremenny' class. Further ships are expected to be built, completing in the late 1980s after trials and evaluation of the 'Kirov' class. These may be significantly larger than their predecessors.

Soviet type name: Atomny raketny kreyser, meaning nuclear-powered missile cruiser.

Frunze *('Kirov' class) (RAAF)*

ЛЕНИНГРАД   113

Number in class: 2

LENINGRAD
MOSKVA

Displacement, tons: 17 500 full load
Dimensions, feet (metres): 620 × 111.5 max
(75.4 wl) × 32.8 (screws) *(189 × 34 (23) × 10)*
Aircraft: 14 Hormone A ASW helicopters
Missiles: SAM; 2 twin SA-N-3 (48 missiles)
Guns: 4—57 mm/70 (twin)
A/S weapons: 1 twin A/S missile launcher (15
SUW-N-1 missiles); 2—12-barrelled RBU 6000
on forecastle
Main engines: Geared steam turbines; 2 shafts;
100 000 shp
Boilers: 4 water tube
Speed, knots: 30
Fuel, tons: 2 600
Range, miles: 9 000 at 18 knots; 4 500 at 29
knots
Complement: 840 plus air wing
Commissioned: 1967-68

This class represented a radical change of
thought in the Soviet fleet. The design must have
been completed while the 'November' class
submarines were building and the heavy A/S
armament and efficient sensors (helicopters and
VDS) suggest an awareness of the problem of
dealing with nuclear submarines. Alongside a
primary A/S role these ships have a capability for
A/A warning and self-defence as well as a
command function. With a full fit of radar and
ECM equipment they clearly represent good
value for money. Why only two were built is
discussed earlier in the notes on the 'Kiev' class
aircraft carriers but it is also significant that the
hull design gave poor performance in heavy
weather.

Soviet type name: Protivolodochny kreyser,
meaning anti-submarine cruiser.

Leningrad *('Moskva' class) 1981*

Number in class: 1 + 2

SLAVA

Displacement, tons: 10 500 standard; 12 500 full load
Dimensions, feet (metres): 613.4 × 65.6 × 25 (187 × 20 × 7.6)
Aircraft: 1 Hormone B helicopter
Missiles: SSM; 16 SS-N-12; SAM; 8 SA-N-6—vertical launchers (64 missiles); 2 SA-N-4 (twin) (40 missiles); ASW; Helicopter carries A/S torpedoes and depth bombs
Guns: 2—130 mm (twin); 6—30 mm Gatlings
A/S weapons: 2 RBU 6000
Torpedo tubes: 8—21 in (533 mm) in hull
Main engines: 4 gas turbines; 2 shafts; 120 000 hp
Speed, knots: 34
Range, miles: 3 500 at 30 knots, 10 000 at 18 knots
Complement: 600 approx
Commissioned: 1982 -

This class is building at the same yard that built the 'Kara' class. First of class started trials in Black Sea in mid-1982.

This is a smaller edition of the dual-purpose surface warfare/ASW *Kirov*, designed as a conventionally powered back-up for that class.

'Slava' class (US Navy, 1983)

Number in class: 7

AZOV
KERCH
NIKOLAYEV
OCHAKOV
PETROPAVLOVSK
TALLINN
TASHKENT

Displacement, tons: 8 200 standard; 9 700 full load

Dimensions, feet (metres): 568 × 61 × 22 *(173.2 × 18.6 × 6.7)*

Aircraft: 1 Hormone A helicopter (hangar aft)

Missiles: A/S; 8—SS-N-14 (2 quad launchers abreast bridge) SAM; 2—SA-N-3 (twin) (72 missiles) (one launcher in *Azov*) SAM; 6—SA-N-6 launchers (*Azov* only)
SAM; 2—SA-N-4 (twin either side of mast) (40 missiles)

Guns: 4—76 mm/59 (2 twin abaft bridge) 4—Gatling 30 mm (abreast funnel)

A/S weapons: 2—12-barrelled RBU 6000 (fwd) 2—6-barrelled RBU 1000 (aft) (not in *Azov* and *Petropavlovsk*)

Torpedo tubes: 10—21 in *(533 mm)* (2 quin mountings abaft funnel) (not in *Azov*)

Main engines: COGOG; 6 gas turbines; 4 of 25 000 shp; 2 of 10 000 shp; 2 shafts

Speed, knots: 34

Range, miles: 1 200 at 15 knots; 3 000 at 32 knots

Complement: 540 (30 officers, 510 ratings)

Commissioned: 1971-79

Apart from the specialised 'Moskva' class this was the first class of large cruisers to join the Soviet Navy since the 'Sverdlov' class and was designed specifically for ASW. *Nikolayev* was first seen in public when she entered the Mediterranean from the Black Sea on 2 March 1973. Clearly capable of prolonged operations overseas. Fitted with stabilisers.
   Azov is of a modified design.

Soviet type name: Bolshoy protivolodochny korabl, meaning large anti-submarine ship.

Left: Tallinn *('Kara' class) (JMSDF 1985)*
Overleaf
Left: Tallinn *('Kara' class) (1985)*
Right: Nikolayev *('Kara' class) (Selçuk Emre)*

Number in class: 10

KRONSHTADT
ADMIRAL ISAKOV
ADMIRAL NAKHIMOV
ADMIRAL MAKAROV
MARSHAL VOROSHILOV
ADMIRAL OKTYABRSKY
ADMIRAL ISACHENKOV
MARSHAL TIMOSHENKO
VASILY CHAPAYEV
ADMIRAL YUMASHEV

Displacement, tons: 6 000 standard; 7 700 full
load
Dimensions, feet (metres): 519.9 × 55.4 × 19.7
(158.5 × 16.9 × 6)
Aircraft: 1 Hormone A helicopter (hangar aft)
Missiles: A/S; 8—SS-N-14 (2 quad launchers)
SAM; 4—SA-N-3 (2 twin launchers) (48
missiles)
Guns: 4—57 mm (2 twin); 4—30 mm Gatling
A/S weapons: 2—12-barrelled RBU 6000 (fwd);
2—6-barrelled RBU 1000 (aft)
Torpedo tubes: 10—21 in *(533 mm)* (2 quin)
Main engines: 2 steam turbines; 2 shafts;
110 000 shp
Boilers: 4 water tube
Speed, knots: 35
Range, miles: 10 500 at 14 knots; 2 400 at 32
knots
Complement: 400
Commissioned: 1969-77

The design was developed from that of the
'Kresta I' class but, with the SS-N-14 replacing
the SS-N-3, the role has been changed from that
of surface warfare for 'Kresta I' to ASW for
'Kresta II'. At the same time the substitution of
SA-N-3 for SA-N-1 improved 'Kresta II' class's air
defence capability. Fin stabilisers. Hulls 8-10
have an additional deck-house abaft the bridge
and between the four Gatling mounts.

Soviet type name: Bolshoy protivolodochny
korabl, meaning large anti-submarine ship.

Left: Admiral Oktyabrsky *('Kresta II'*
class)
Overleaf
Left: Admiral Yumashev *('Kresta II'*
class) (1983)
Right: *'Kresta II' class*

Number in class: 4

ADMIRAL ZOZULYA
VLADIVOSTOK
VITSE-ADMIRAL DROZD
SEVASTOPOL

Displacement, tons: 6 140 standard; 7 600 full load
Dimensions, feet (metres): 510 × 55.7 × 19.7 *(155.5 × 17 × 6)*
Aircraft: 1 Hormone B helicopter with hangar aft
Missiles: SSM; 4—SS-N-3B (2 twin launchers) (no reloads); SAM; 4—SA-N-1 (2 twin launchers) (32 missiles)
Guns: 4—57 mm (twin); 4—30 mm Gatling *(Drozd only)*
A/S weapons: 2—12-barrelled RBU 6000 (fwd); 2—6-barrelled RBU 1000 (aft)
Torpedo tubes: 10—21 in *(533 mm)* (2 quin)
Main engines: Steam turbines; 2 shafts; 110 000 shp
Boilers: 4 water tube
Speed, knots: 32
Range, miles: 8 000 at 14 knots; 2 400 at 32 knots
Complement: 375
Commissioned: 1967-69

Designed for surface warfare, the successors to the 'Kynda' class.

Provided with a helicopter landing deck and hangar aft—the latter for the first time in a Soviet ship. This gives an enhanced carried-on-board target-location facility for the 120 n mile SS-N-3B system. The 'Kresta I' was therefore the first Soviet missile cruiser free to operate alone and without targeting assistance from own shore-based aircraft.

The first of class, *Admiral Zozulya*, carried out sea trials in the Baltic in February 1967.

Soviet type name: Originally bolshoy protivolodochny korabl, meaning large anti-submarine ship. Changed in 1977-78 to raketny kreyser, meaning missile cruiser.

'Kresta I' class

Number in class: 4

**GROZNY
ADMIRAL FOKIN
ADMIRAL GOLOVKO
VARYAG**

Displacement, tons: 4 400 standard; 5 550 full load
Dimensions, feet (metres): 465.8 × 51.8 × 17.4 *(142 × 15.8 × 5.3)*
Aircraft: Pad for helicopter on stern
Missiles: SSM; 8—SS-N-3B (quad launchers with one reload per tube); SAM; 2—SA-N-1 (twin launcher) (16 missiles)
Guns: 4—76 mm (twin); 4—30 mm Gatlings
A/S weapons: 2—12-barrelled RBU 6000 on fo'c'sle
Torpedo tubes: 6—21 in *(533 mm)* (2 triple amidships)
Main engines: 2 sets geared steam turbines; 2 shafts; 90 000 shp
Boilers: 4 high pressure
Speed, knots: 34
Range, miles: 6 000 at 14.5 knots; 1 500 at 34 knots
Complement: 390
Commissioned: 1962-65

This class was designed for surface warfare and was the first class of missile cruisers built. The role made it the successor of the 'Sverdlov' class. Two enclosed towers, instead of masts, are stepped forward of each raked funnel. In this class there is no helicopter embarked, so guidance for the SS-N-3B system would be more difficult than in later ships. They are therefore constrained in their operations compared with the later ships with their own helicopters. Three of this class have undergone modernisation, including the installation of four 30 mm Gatling mounts with two assocaited Bass Tilt radars. *Admiral Golovko* was the last to be taken in hand in 1985-86.

Soviet type name: Raketny kreyser, meaning missile cruiser.

Admiral Fokin *('Kynda' class) 1980*
(US Navy)

Number in classes: 1 + 2 + 9

ADMIRAL SENYAVIN (CG)
ADMIRAL USHAKOV
ALEKSANDR NEVSKY
ALEKSANDR SUVOROV
DMITRY POZHARSKY
DZERZHINSKY (CG)
MIKHAIL KUTUZOV
MURMANSK
OKTYABRSKAYA REVOLUTSIYA
SVERDLOV
ZHDANOV (CG)

Left: Zhdanov ('Sverdlov' class) (Royal
Navy, 1984)
Overleaf
Left: Murmansk ('Sverdlov' class)
(Royal Danish Navy, 1980)
Right: Oktyabrskaya Revolutsiya
('Sverdlov' class) (MoD, 1981)

Displacement, tons: 16 000 standard; 17 200 full load
Dimensions, feet (metres): 689 × 72.2 × 23.6 (210 × 22 × 7.2)
Aircraft: Helicopter pad in Zhdanov. Pad and hangar for Hormone helicopter in Senyavin
Armour: Belts 3.9—4.9 in (100-125 mm); fwd and aft 1.6—2 in (40-50 mm); turrets 4.9 in (125 mm); C.T. 5.9 in (150 mm); decks 1—2 in (25-50 mm) and 2—3 in (50-75 mm)
Missiles: SAM; 2—SA-N-2 (twin launcher) aft in Dzerzhinsky;SAM; 2—SA-N-4 in Zhdanov and Senyavin (twin launcher) (20 missiles) (see Conversions)
Guns: 12—152 mm (4 triple), 12—100 mm (6 twin); 32—37 mm (twin); 16—30 mm (twin) (in Oktyabrskaya Revolutsiya, Admiral Ushakov and Aleksandr Suvorov) Dzerzhinsky; 9—152 mm (3 triple); 12—100 mm (6 twin); 16—37 mm (twin) A. Senyavin; 6—152 mm (2 triple); 12—100 mm (6 twin); 32—37 mm (twin); 16—30 mm (twin) Zhdanov; 9—152 mm (3 triple); 12—100 mm (6 twin); 32—37 mm (twin); 8—30 mm (twin) (see Conversions note)
Mines: 150 capacity—(except Zhdanov and Senyavin)
Main engines: Geared turbines; 2 shafts; 110 000 shp
Boilers: 6 water tube
Speed, knots: 32
Oil fuel, tons: 3 800
Range, miles: 8 700 at 18 knots; 2 700 at 32 knots

Complement: 1 000 average

Of the 24 cruisers of this class originally projected, 20 keels were laid at Leningrad, Nikolayev and Severodvinsk and 17 hulls were launched from 1951 onwards, but only 14 ships were completed by 1956. There were two slightly different types. Sverdlov and sisters had the 37 mm guns near the fore-funnel one deck higher than in later cruisers. All ships except Zhdanov and Senyavin are fitted for minelaying. Mine stowage is on the second deck. Sverdlov, Admiral Lazarev and Dzerzhinsky in reserve.

Conversions: Dzerzhinsky was fitted with an SA-N-2 launcher aft replacing X-turret in 1961— presumably the experiment was not sufficiently successful to extend this item. In 1972 Admiral Senyavin returned to service with both X and Y turrets removed and replaced by a helicopter pad and a hangar surmounted by four 30 mm mountings and an SA-N-4 mounting. At about the same time Zhdanov had only X-turret removed and replaced by a high deckhouse mounting an SA-N-4. Both conversions took about three years.
Oktyabrskaya Revolutsiya completed refit in early 1977 and Admiral Ushakov and Aleksandr Suvorov in 1979 which included the extension of bridge work aft, the fitting of eight 30 mm mounts with associated four radars and the removal of radars from 100 mm turrets.

Number in class: 5 + 4

SOVREMENNY
OTCHYANNY
OTLICHNNY
OSMOTRITELNY
BEZUPRECHNY

Displacement, tons: 6 000 standard; 7 900 full load
Dimensions, feet (metres): 511.8 × 56.8 × 21.3 *(156 × 17.3 × 6.5)*
Aircraft: 1 Helix helicopter
Missiles: SSM; 8—SS-N-22 (quad either side of bridge); SAM; 2—SA-N-7 (one abaft A turret—one between Y turret and helicopter deck—44 missiles)
Guns: 4—130 mm (twin turrets); 4—30 mm Gatlings
Torpedo tubes: 4—21 in *(533 mm)* (twin)
A/S weapons: 2—6-barrelled RBU 1000 (aft) (120 rockets)
Mines: Have minerals
Main engines: 2 steam turbines; 2 shafts; 110 000 shp
Speed, knots: 32
Range, miles: 2 000 at 32 knots; 5 000 at 20 knots
Complement: 320
Commissioned: 1980 -?

Built at the same yard as the 'Kresta II' class. In August 1980 *Sovremenny* appeared in the Baltic without a weapon fit, apparently on engine trials. Missiles and guns were fitted during the winter 1980-81. She carries a helicopter in a telescopic hangar, the aircraft probably being for target acquisition for the SSMs. This class is purpose-built for surface warfare being complemented by the ASW capable 'Udaloy' class.

Class: *Sovremenny* transferred to the Black Sea Fleet in January 1982 probably for trials of her missile systems and was then transferred to Northern Fleet August 1982. *Otchyanny* started trials in Baltic April 1982 and transferred to Northern Fleet October 1982.
  Otlichnny began trials in the Baltic in May 1983, being transferred later in 1984. *Osmotritelny* began sea trials in the Baltic in June 1984 and transferred to the Pacific in August-September 1985.

Soviet type name: Eskadrenny minonosets, meaning destroyer.

Osmotritelny *('Sovremenny' class) (RAAF, 1984)*

Number in class: 7 + 2

UDALOY
VITSE-ADMIRAL KULAKOV
MARSHAL VASILEVSKY
ADMIRAL ZAKOROV
ADMIRAL SPIRIDONOV
ADMIRAL TRIBUTS
MARSHAL SHAPOSHNIKOV

Displacement, tons: 8 000 full load
Dimensions, feet (metres): 537.9 × 61.7 × 20.3
 (164 × 18.8 × 6.2)
Aircraft: 2 Helix A helicopters (2 hangars)
Missiles: ASW; 8 SS-N-14 (quad); SAM; SA-N-8
 PDMS, 8 vertical launchers (64 missiles)
Guns: 2—100 mm (single, fwd); 4—30 mm
 Gatlings
A/S weapons: 2 RBU 6000 (12-barrelled)
Torpedo tubes: 8—21 in *(533 mm)* (quad)
Mines: Have minerails
Main engines: COGOG; 4 gas turbines; 2 shafts;
 110 000 shp total
Speed, knots: 32
Range, miles: 2 000 at 32 knots; 6 000 at 20
 knots
Complement: 300
Commissioned: 1980-87?

A new design, originally known as BAL COM 3 in
NATO, successor to 'Kresta II' class designed for
ASW and complementary to the 'Sovremenny'
class. Has four funnels similar to 'Kashin' class.
Udaloy carried out trials in the Baltic from late
1980, *Vitse-Admiral Kulakov* started trials in
September 1981. First two now in Northern
Fleet.
 Marshal Vasilevsky started trials in the Baltic in
July 1983 and transferred to the Northern Fleet
in April 1984. *Admiral Zakorov* started trials in
October 1983, *Admiral Spiridonov* in September
1984. Both transferred to Pacific in August-
September 1985.

Soviet type name: Bolshoy protivolodochny
korabl, meaning large anti-submarine ship.

Udaloy *('Udaloy' class) 1983 (DoD)*

**Number in classes: 13 + 6
(Conversions)**

KOMSOMOLETS UKRAINY
KRASNY KAVKAZ
KRASNY KRIM
OBRAZTSOVY (**)
ODARENNY (**)
OGNEVOY* (**)
PROVORNY (see Conversion note)
RESHITELNY
SDERZHANNY*
SKORY
SLAVNY* (**)
SMELY*
SMETLIVY
SMYSHLENNY*
SOOBRAZITELNY
SPOSOBNY
STEREGUSHCHY (**)
STROGY
STROYNY*

* modified (see Conversion note)
(**) Zhdanov-built, remainder at Nikolayev

Left: 'Kashin' class (1972)
Overleaf
Left: Stroyny ('Mod. Kashin' class)
(RAF, 1983)
Right: Strogy ('Mod. Kashin' class)
(RAAF, 1984)

Displacement, tons: 3 750 standard; 4 500 full
load ('Kashin') 3 950/4 900 (mod 'Kashin')
Dimensions, feet (metres): 472.4 (482.3 mod) ×
51.8 × 15.4 *(144 (147) × 15.8 × 4.7)*
Aircraft: Helicopter landing deck
Missiles: SSM; 4—SS-N-2C (single) (mod
'Kashin') (no reloads); SAM; 4—SA-N-1 (twin
launchers) (32 missiles) (except *Provorny*, see
note)
Guns: 4—76 mm (twin); 4—30 mm Gatlings
(mod 'Kashin' only)
A/S weapons: 2—12-barrelled RBU 6000 fwd;
2—6-barrelled RBU 1000 aft (unmodified only)
Torpedo tubes: 5—21 in *(533 mm)* (quin)
amidships
Mines: Laying capability
Main engines: 4 sets gas turbines; 2 shafts;
94 000 hp
Speed, knots: 36
Range, miles: 2 500 at 25 knots; 1 500 at 35
knots
Complement: 280
Commissioned: 1963-72

The first class of warships in the world to rely
entirely on gas-turbine propulsion. These ships
were delivered from 1963 to 1972—from the
Zhdanov Yard, Leningrad (1964-66) and from the
61 Kommuna (North) Yard, Nikolayev (1963-72).
Sderzhanny, last of the class, was the only one to

be built to the modified design—the other five
were converted after completion.

As they were built at the same time as 'Kynda'
class they may originally have been intended as
AA support for the latter.

Soviet type name: Bolshoy protivolodochny
korabl, meaning large anti-submarine ship.

Conversion: In order to bring this class up-to-
date with SSM and VDS a conversion programme
was started in 1972 to the same pattern as set in
Sderzhanny. This conversion consists of
lengthening the hull by 10 ft *(3 m)*, shipping four
SS-N-2 (C) launchers (SSM), four Gatling close
range weapons, a VDS under a new stern
helicopter platform and removing the after RBUs.
By 1976 five had been so converted. The sixth
and last conversion was completed in 1980.

In the mid-1970s *Provorny* was converted to
carry out trials of the new SA-N-7 SAM system
(VLS) for new construction ships. After lengthy
trials in the Black Sea she was transferred to the
Northern Fleet for eight months in 1981 before
returning to the Black Sea in May 1982.

Transfers: Additional ships of a modified design
are being built for India. First transferred
September 1980, the second in June 1982, the
third in 1983 and the fourth in July 1986.

Number in classes: 3 + 1

BEDOVY
NEULOVIMY
NEUDERZHIMY
PROZORLIVY (mod)

Prozorlivy *('Mod. Kildin' class) 1980*

Displacement, tons: 3 000 standard; 3 500 full load

Dimensions, feet (metres): 414.9 × 42.6 × 15.1 *(126.5 × 13 × 4.6)*

Missiles: SSM; 4—SS-N-2C (single) (modified ships)

Guns: 4—76 mm (twin aft) (after modification); 16—57 mm (quad—2 fwd, 2 between funnels) (45 mm in *Bedovy*)

A/S weapons: 2—16-barrelled RBU 2500 on fo'c'sle

Torpedo tubes: 4—21 in *(533 mm)* (2 twin)

Main engines: Geared turbines; 2 shafts; 72 000 shp

Boilers: 4 high pressure

Speed, knots: 36

Range, miles: 3 500 at 18 knots; 1 100 at 34 knots

Complement: 300 officers and men

Commissioned: 1957-58

The first ship of this class, *Bedovy,* from the shape of her funnels was altered from the 'Kotlin' DD design to 'Kildin' DDG during construction, being followed by the other three ships of this class built from the keel up as 'Kildin' class. However the design was based on the 'Kotlin' hull with the SS-N-1 replacing the after 130 mm turret and with two quadruple 57 mm mountings in place of the forward 130 mm turret. *Bedovy* built at Nikolayev, *Neuderzhimy* at Komsomolsk,

Neulovimy at Leningrad and *Prozorlivy* at Nikolayev. All completed 1957-58. In 1971 *Neulovimy* was taken in hand for modification. This was completed by 1973 and consisted of the replacement of the SS-N-1 on the quarterdeck by two superimposed twin 76 mm turrets, the fitting of four SS-N-2C launchers abreast the after funnel and the fitting of new radar. The substitution of the 40 n mile SS-N-2C system (a modified Styx) for the obsolescent SS-N-1 system and the notable increase in gun armament illustrate two trends in Soviet thought. *Bedovy* and *Prozorlivy* completed this conversion in 1973 and 1976. The fourth ship *Neuderzhimy* is probably unlikely to be modernised as she is now more than 25 years old. Her SS-N-1 system has probably been removed.

Soviet type name: Originally bolshoy protivolodochny korabl, meaning large anti-submarine ship. Changed in 1977-78 to bolshoy raketny korabl, meaning large missile ship, the modifications having given this class a surface warfare role (except for *Neuderzhimy*).

Number in class: 8

BOYKY
DERZKY
GNEVNY
GORDY
GREMYASHCHY
UPORNY
ZORKY
ZHGUCHY

Displacement, tons: 3 700 standard; 4 750 full load
Dimensions, feet (metres): 455.9 × 48.2 × 16.4 *(139 × 14.7 × 5)*
Aircraft: Helicopter platform
Missiles: SAM; 2—SA-N-1 (twin launcher aft) (16 missiles)
Guns: 8—57 mm (2 quad fwd); 8—30 mm (twin) (by after funnel)
A/S weapons: 3—12-barrelled RBU 6000
Torpedo tubes: 10—21 in *(533 mm)* (2 quin)
Main engines: 2 sets geared steam turbines; 2 shafts; 85 000 shp
Boilers: 4 water tube
Speed, knots: 35
Oil fuel, tons: 900
Range, miles: 4 500 at 18 knots; 1 100 at 32 knots
Complement: 350

Five ships of this class were converted from 'Krupny' class at Zhdanov Yard, Leningrad (1968-72) and three in the Pacific (1974-78), being given a SAM capability instead of the latter's SSM armament. The original 'Krupny' class was built at Leningrad, Nikolayev and in the Pacific between 1957 and 1962.

Soviet type name: Bolshoy protivolodochny korabl, meaning large anti-submarine ship.

Zhguchy ('Kanin' class) 1981

Number in class: 8

BRAVY*
NAKHODCHIVY
NASTOYCHIVY
NESOKRUSHIMY**
SKROMNY
SKRYTNY**
SOZNATELNY**
VOZBUZHDENNY**

* Original R and D conversion
** Modified guns (30 mm)

Displacement, tons: 2 850 standard; 3 500 full load
Dimensions, feet (metres): 414.9 × 42.3 × 15.1 *(126.5 × 12.9 × 4.6)*
Missiles: SAM; SA-N-1 (twin launcher) (16 missiles)
Guns: 2—130 mm (twin); 4—45 mm (quad); (12—45 mm in *Bravy*); 8—30 mm (twin) (in four ships)
Torpedo tubes: 5—21 in *(533 mm)* (quin)
A/S weapons: 2—12-barrelled RBU 6000 (2—16-barrelled RBU 2500 in *Bravy* and *Skromny*)
Main engines: Geared turbines; 2 shafts; 72 000 shp
Boilers: 4 high pressure
Speed, knots: 36
Oil fuel, tons: 800
Range, miles: 4 500 at 12 knots; 1 100 at 34 knots
Complement: 300

Converted 'Kotlin' class destroyers with a surface-to-air missile launcher in place of the main twin turret aft and anti-aircraft guns reduced to one quadruple mounting in all but *Bravy* which still has three quadruple mountings. The prototype *(Bravy)* was completed in 1961 and the others between 1967 and 1972. Four subsequently modified with 30 mm armament and different radar.

Appearance: The prototype SAM 'Kotlin' *(Bravy)* has a different after funnel and different radar pedestal from those in the rest of the class.

Soviet type name: Eskadrenny minonosets (esminets), meaning destroyer.

Transfer: One to Poland—ex-*Spravedlivy* in 1970. Renamed *Warszawa*, deleted 31 Dec 1985.

Vozbuzhdenny *('SAM Kotlin' class)*
(US Navy, 1980)

Number in classes: 12 + 6

BLAGORODNY*
BLESTYASHCHY*
BURLIVY*
BYVALY*
DALNEVOSTOCHNY KOMSOMOLETS*
MOSKOVSKY KOMSOMOLETS*
NAPORISTY*
PLAMENNY*
SPESHNY
SPOKOYNY
SVEDUSHCHY*
SVETLY
VDOKHNOVENNY*
VESKY
VLIYATELNY
VOZMUSHCHENNY*
VYDERZHANNY*
VYZYVAYUSHCHY*

* Modified ships

Displacement, tons: 2 850 standard; 3 500 full load
Dimensions, feet (metres): 414.9 × 42.3 × 15.1 *(126.5 × 12.9 × 4.6)*
Guns: 4—130 mm (2 twin); 16—45 mm (4 quad); 4—25 mm (twin); 4 or 8—25 mm (twin) (in 'Modified Kotlin' class)
A/S weapons: 2—16-barrelled RBU 2500 (fwd); 2—6-barrelled RBU 600 (aft) ('Modified Kotlin' class—*M. Komsomolets* has two RBU 6000 forward—no weapons aft; 6 DCT; 2 DC racks (unmod except *Svetly* and *Vesky* which have a helo platform)
Torpedo tubes: 5 (mod) and 10 (unmod) 21 in *(533 mm)* (quin)
Mines: 56 capacity
Main engines: Geared turbines; 2 shafts; 72 000 shp
Boilers: 4 high pressure
Speed, knots: 36
Range, miles: 4 500 at 12 knots; 1 100 at 34 knots
Complement: 285

Built in 1954-56. Probably the last hull laid down was completed as 'Kildin' class. Twelve modified in early 1960s. Four 'Kotlin' in reserve.

Modifications: (a) Eight converted to 'SAM Kotlin' class plus one transferred to Poland. (b) *Svetly* only ship now provided with helicopter platform on stern. (c) Modified ships had after torpedo tubes replaced by a deckhouse. (d) Modified ships had two 16-barrelled RBUs (forward) and two 6- barrelled RBUs (aft) fitted. (e) The latest addition in some ships is the fitting of four or eight 25 mm either side of the after-funnel in place of two quad 45 mm.

Soviet type name: Eskadrenny minonosets (esminets), meaning destroyer.

Left: Vliyatelny *('Kotlin' class) (US Navy, 1975)*
Overleaf
Left: Speshny *('Kotlin' class) 1978*
Right: Blagorodny *('Mod. Kotlin' class) 1982*

Number in class: 6 (+ 6 Reserve)

Displacement, tons: 2 240 standard; 3 130 full load

Dimensions, feet (metres): 395.2 × 38.9 × 15.1 *(120.5 × 11.9 × 4.6)*

Guns: Unmodified: 4—130 mm/50 (twin); 2—85 mm (twin); 2—57 mm (twin); 8—37 mm (4 twin) or 7—37 mm (single); 4 or 6—25 mm (twin) Modified: 4—130 mm/50 (twin) 5—57 mm (single)

A/S weapons: 2 internal DC racks (unmodified); 2—16-barrelled RBU 2500 (modified)

Torpedo tubes: 10—21 in *(533 mm)* (unmodified); 5—21 in *(533 mm)* (modified)

Mines: 50 can be carried

Main engines: 2 geared turbines; 2 shafts; 60 000 shp

Boilers: 4 high pressure

Speed, knots: 33

Range, miles: 3 500 at 14 knots; 940 at 30 knots

Complement: 280

There were to have been 85 destroyers of this class, but construction beyond 75 was discontinued in favour of later types of destroyers, and the number has been further reduced to 12 by transfers to other countries, translations to other types and disposals. Further deletions expected in this elderly class of which six are in reserve. All originally built between 1949 and 1954 at Nikolayev, Severodvinsk and Leningrad.

Appearance: There were three differing types in this class, the anti-aircraft guns varying with twin and single mountings; and two types of foremast, one vertical with all aerials on top and the other with one aerial on top and one on a platform half way.

Soviet type name: Eskadrenny minonosets (esminets), meaning destroyer.

Modernisation: About eight ships of the 'Skory' class were modified between 1958 and 1960 including extensive alterations to anti-aircraft armament, electronic equipment and anti-submarine weapons. These now have five 57 mm (single) in place of the 85 mm and 37 mm, five torpedo tubes and two 16- barrelled RBUs. Five remain in service in the Soviet Navy.

'Skory' class (1976)

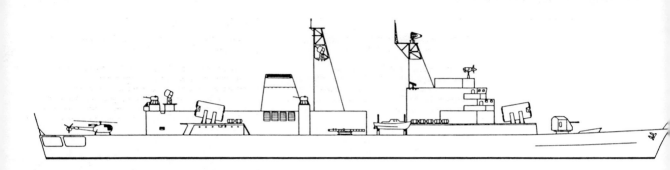

Displacement, tons: 5-6000 approx
Dimensions, feet (metres): 475.6 × 52.5 × —
  (145 × 16 × —)
Aircraft: Hangar for 2 helicopters
Missiles: SSM; 8—SS-N-2C; SAM; ? SA-N-4
Guns: 1 or 2—76 mm; 6—30 mm (twins)
Torpedo tubes: 2—21 in (533 mm) mountings
Main engines: Gas turbines
Speed, knots: 28

Built at Mangalia, launched 1982, commissioned 5 August 1985.

Armament: The SSMs are mounted as follows—two twin in B position with a possible SA-N-4 in between and a further pair either side of the hangar. The gun armament consists of a single or twin turret on forecastle and three mounts, either 30 mm (twin) or Gatling on each side of funnel and on hangar.

Muntenia. *(Provisional drawing copyright JDW)*

Number in classes: 21 + 11 + 2 (+ 2)

'KRIVAK I'

BDITELNY
BEZUKORIZNENNY
BEZZAVETNY
BODRY
DEYATELNY
DOBLESTNY
DOSTOYNY
DRUZHNY
LADNY
LENINGRADSKY KOMSOMOLETS*
LETUCHY*
PORYVISTY
PYLKY*
RAZUMNY
RAZYASHCHY
RETIVY*
SILNY
STOROZHEVOY
SVIREPY
ZADORNY*
ZHARKY*

'KRIVAK II'

BESSMENNY
GORDELIVY

Left: *Krivak I' class 1980 (G. Jacobs)*
Overleaf
Left: Razitelny *('Krivak II' class) (Royal Navy, 1979)*
Right: Menzhinsky *('Krivak III' class) (JMSDF, 1984)*

GROMKY
GROZYASHCHY
NEUKROTIMY
PYTLIVY
RAZITELNY
REVNOSTNY
REZKY
REZVY
RYANNY

'KRIVAK III'

MENZHINSKY
DZERZHINSKY

* Modified ships

Displacement, tons: 3 100 standard; 3 900 full load
Dimensions, feet (metres): 405.2 × 45.9 × 16.4 *(123.5 × 14 × 5)*
Aircraft: See note
Missiles: A/S; 4—SS-N-14 in A position (quad launcher) (not in 'Krivak III'); SAM; 4—SA-N-4 (twin launchers) (40 missiles); 2 (twin) in 'Krivak III' (20 missiles)
Guns: 4—76 mm (2 twin) X and Y positions in 'Krivak I'; 2—100 mm (single, aft) in 'Krivak II'; 1—100 mm (fwd), 2—30 mm Gatling in 'Krivak III'
A/S weapons: 2—12-barrelled RBU 6000 (fwd)
Torpedo tubes: 8—21 in *(533 mm)* (2 quad)
Mines: Capacity for 20

Main engines: COGAG; 4 gas turbines (2 cruising, 14 000 shp; 2 boost, 55 000 shp); 2 shafts; 69 000 shp
Speed, knots: 32
Range, miles: 4 600 at 20 knots; 1 600 at 30 knots
Complement: 180

This handsome class, the first ship of which appeared in 1970, incorporates A/S and anti-air capability, a VDS, RBUs, two banks of tubes, all in a hull designed for both speed and sea-keeping. The use of gas turbines gives the 'Krivak' class a rapid acceleration and availability. Classified as 'missile frigates' by NATO.

'Krivak III' class is being built for the KGB. The removal of SS-N-14 compensates for the addition of a hangar and flight deck for a Hormone or Helix helicopter. *Menzhinsky* deployed from Black Sea to Pacific in September 1984 and *Dzerzhinsky* in September 1985.

Class: 'Krivak II' class has X-gun mounted higher and the break to the quarter-deck further aft apart from other variations noted.

Soviet type name: Originally bolshoy protivolodochny korabl, meaning large anti-submarine ship. Changed in 1977-78 to storozhevoy korabl, meaning escort ship.

Number in class: 1 (+ 3)

USSR
DELFIN

GERMANY (DR)
BERLIN
ROSTOCK
+ 1

Displacement, tons: 1 700 standard; 1 900 full load

Dimensions, feet (metres): 311.6 × 42 × 13.7 *(95 × 12.8 × 4.2)*

Missiles: SAM; 2—SA-N-4 (twin launcher) (20 missiles)

Guns: 4—3 in *(76 mm)* (twin); 4—30 mm (twin)

A/S weapons: 2—12-barrelled RBU 6000 (fwd); 2 DC racks

Mines: Capacity for 20

Main engines: CODAG; 1 gas turbine (centre shaft); 18 000 shp; 2 diesels (outer shafts); 12 000 shp

Speed, knots: 27 (gas); 22 (diesel)

Range, miles: 1 800 at 14 knots

Complement: 110

First reported in the Black Sea in 1976. This ship has been retained by the USSR for training foreign crews—all other new construction has been exported. Of the three 'Koni' class destroyers in service with the German Democratic Republic Navy two — *Rostock* and *Berlin* — were commissioned in 1978 and 1979; the third was commissioned at Leningrad early in 1986. *Rostock* and *Berlin* differ from the basic 'Koni' class in having a maximum speed of 32 knots (gas) and 22 knots (diesel) and a complement of 130.

Class: There are two types of this class, Type II having the space between the funnel and the after superstructure filled in with an extra deckhouse which may contain air- conditioning machinery for warmer areas.
   Twelfth of class building.

Soviet type name: Strozhevoy korabl, meaning escort ship.

Berlin *('Koni' class) of the German Democratic Republic (RAF, 1981)*

Number in class: 45

USSR

ARKHANGELSKY KOMSOMOLETS
ASTRAKHANSKY KOMSOMOLETS
BARS
BARSUK
BOBR
KOBCHIK
KOMSOMOLETS GRUZY
KOMSOMOLETS LITVY
KRASNODARSKY KOMSOMOLETS
KUNITSA
LEOPARD
PANTERA
ROSOMOKHA
RYS
SOVETSKY AZERBAYDZHAN
SOVETSKY DAGESTAN
SOVETSKY TURKMENISTAN
TUMAN
VOLK
VORON
+25

BULGARIA

DRUZKI
SMELI (+1)

Displacement, tons: 1 000 standard; 1 510 full load
Dimensions, feet (metres): 300.1 × 33.1 × 10.5 *(91.5 × 10.1 × 3.2)*
Guns: 3—100 mm (single); 4—37 mm (twin); 4—25 mm (twin) in some
A/S weapons: 2—16-barrelled RBU 2500; 2 DC racks, 4 DCTs (some)
Torpedo tubes: 2 or 3—21 in *(533 mm)* (twin or triple)
Mines: 28
Main engines: 2 geared turbines; 2 shafts; 20 000 shp
Boilers: 2
Speed, knots: 30
Range, miles: 2 000 at 13 knots; 700 at 27 knots
Complement: 175

Built from 1952 to 1959 at several yards. Successors to the 'Kola' class escorts (now deleted) of which they are lighter and less heavily armed but improved versions. About ten in reserve.

Conversion: A small number of this class has been improved. Some have a twin 25 mm gun mounting on either side of the funnel and a dipping sonar abreast of the bridge.

Soviet type name: Storozhevoy korabl, meaning escort ship.

Transfers: Two to Bulgaria (1957-58), four to German Democratic Republic (1956-59), two to Finland (1964), eight to Indonesia (1962-65). Many of these have been scrapped. The first two Bulgarian 'Riga' class ships were extensively refitted 1980-81. A third (ex-*Kobchik*) may have been transferred in November 1985.

Druzki *('Riga' class) of the Bulgarian Navy 1984 (Selçuk Emre)*

Number in classes: 15 + 11 + 32 + 2

AMETYST
BRILLIANT
IZUMRUD
RUBIN
SAPFIR
ZHEMCHUG
(All 'Grisha II' class (KGB))

Displacement, tons: 950 standard; 1 200 full load
Dimensions, feet (metres): 236.2 × 32.8 × 12.1 *(72 × 10 × 3.7)*
Missiles: SAM; 2—SA-N-4 (twin launcher) ('Grisha I' 'III' and 'V' classes) (20 missiles)
Guns: 2—57 mm (twin) (4 (2 twin) in 'Grisha II' class); 1—76 mm ('Grisha V'); 30 mm Gatling mount aft ('Grisha III' and 'V' classes)
Torpedo tubes: 4—21 in *(533 mm)* (2 twin)
A/S weapons: 2 RBU 6000 (12-barrelled); 2 DC racks
Mines: Capacity for 18
Main engines: CODAG; 1 gas turbine; 15 000 shp; 2 diesels; 16 000 shp; 3 shafts $_5$ 30 knots
Range, miles: 4 500 at 10 knots; 1 750 at 22 knots (diesels); 950 at 27 knots
Complement: 70 (III); 60 (I)
Commissioned: 1968-

'Grisha I' started series production in late 1968. Continuing programme of about three a year. SA-N-4 launcher mounted on the fo'c'sle in all but 'Grisha II'. This is replaced by a second twin 57 mm in 'Grisha II' class which are operated by KGB. 'Grisha III' has improved radar and Gatling fitted aft. 'Grisha V' is similar to 'Grisha III' with the after twin 57 mm mounting replaced by a single 'Tarantul' Type 76 mm gun.

Soviet type name: Maly protivolodochny korabl, meaning small anti-submarine ship ('Grisha I and III and V') or pogranichny storozhevoy korabl, meaning border patrol ship ('Grisha II'). Only the KGB ships have names.

Left: *'Grisha III' class*
Overleaf
Left: *'Grisha I' class*
Right: *'Grisha II' class*

Number in classes: 9 + 9

Displacement, tons: 950 standard; 1 150 full
   load
Dimensions, feet (metres): 270.3 × 29.9 × 9.8
   (82.4 × 9.1 × 3)
Guns: 4—76 mm (twin)
A/S weapons: 4—RBU 6000 (2 fwd, 2 aft)
   ('Mirka I') 2—RBU 6000 (fwd) ('Mirka II'); DC
   rack in some 'Mirka I'
Torpedo tubes: 5—16 in *(406 mm)* ('Mirka I')
   (quin); 10—16 in *(406 mm)* ('Mirka II') (quin)
Main engines: CODAG; 2 diesels; 12 000 bhp; 2
   gas turbines; 25 000 hp; 2 shafts
Speed, knots: 32
Range, miles: 4 800 at 10 knots; 500 at 30 knots
Complement: 96

This class of ships was built at Kaliningrad in
1964-67 as variation on 'Petya' class. The
difference between the Mk I and II is that the
latter have the after RBU rocket launchers
removed and an additional quintuple 16 in
(406 mm) torpedo mounting fitted between the
bridge and the mast.

Soviet type name: Originally maly
protivolodochny korabl, meaning small anti-
submarine ship. Changed in 1977-78 to
storozhevoy korabl, meaning escort ship.

'Mirka II' class (1979)

Number in classes: 7 + 11 + 21 + 1

Displacement, tons: 950 standard; 1 180 full load
Dimensions, feet (metres): 268.3 (270.6, Mod II) × 29.9 × 9.5 *(81.8 (82.5, Mod II) × 9.1 × 2.9)*
Guns: 4—76 mm (twin); (2—76 mm (twin) in some 'Mod Petya I')
A/S weapons: 4—RBU 2500 ('Petya I'); 2—RBU 2500 ('Mod Petya I'); 2—RBU 6000 ('Petya II' and 'Mod Petya II'); 2 internal DC racks in all 'Petya I' and 'Petya II' and 1 in some 'Mod Petya I'
Torpedo tubes: 5—16 in *(406 mm)* (quin) ('Petya I' and 'Mod Petya I' (except for R and D ship with none) and 'Mod Petya II'); 10—16 in *(406 mm)* (2 quin) ('Petya II')
Mines: Capacity for 22 in all but 'Mod Petya I'
Main engines: CODAG; 1 diesel; 6 000 hp; 2 gas turbines; 30 000 hp; 3 shafts
Speed, knots: 32
Range, miles: 4 870 at 10 knots; 450 at 29 knots
Complement: 98

Small freeboard with a low wide funnel. The first ship reported to have been built in 1960-61 at Kaliningrad. Construction continued there and at Komsomolsk until about 1970. 'Petya II' class mounts an extra quintuple torpedo tube in place of after RBUs in 'Petya I'. At least 20 export versions transferred to other countries.

Class: 'Petya I'—the basic hull and armament. 'Mod Petya I'—carry an MF towed sonar in a deckhouse on the stern. One has a large towed sonar in the open with no deckhouse. Another has a deckhouse abaft the funnel replacing the quintuple torpedo tube mounting and a reel and winch on the stern similar to a towed array, while a third has a deckhouse smaller than the remainder on the stern.
'Petya II'—basic hull and armament.
'Mod Petya II'—first seen in 1978. After torpedo tubes removed and with a deckhouse built abaft the after 76 mm gun to contain the towed sonar gear. There is a space either side for mine rails.

Soviet type name: Originally maly protivolodochny korabl, meaning small anti-submarine ship. Changed in 1977-78 to storozhevoy korabl, meaning escort ship.

Left: *'Petya I' class*
Overleaf
Left: *'Petya II' class (US Navy)*
Right: *'Petya II' class (1980)*

Number in class: 17 + 9 (+ 1)

Displacement, tons: 560 standard; 660 full load
Dimensions, feet (metres): 194.5 × 42.6 × 8.5
   (59.3 × 13 × 2.6)
Missiles: SSM; 6—SS-N-9 (2 triple launchers);
   SAM; 2—SA-N-4 (twin launcher) (20 missiles)
Guns: 2—57 mm/L70 (twin) ('Nanuchka I'); 1—
   76 mm/L59; 1—30 mm Gatling ('Nanuchka III')
Main engines: 6—M 504 diesels; 3 shafts;
   30 000 bhp
Speed, knots: 36
Range, miles: 2 500 at 12 knots; 900 at 31 knots
Complement: 60

Probably mainly intended for deployment in coastal waters although occasionally deployed in the Mediterranean (in groups of two or three), North Sea and Pacific. Built from 1969 onwards. Built at Petrovsky, Leningrad and in the Pacific ('Nanuchka III' only). SS-N-9 has a range of about 60 n miles.

'Nanuchka III', first seen in 1978, has a 76 mm gun in place of the twin 57 mm and an added Gatling as in 'Grisha III', different radar and a heavier mast structure.

Countermeasures: Two receivers and two chaff launchers.

Soviet type name: Maly raketny korabl, meaning small missile ship.

Transfers: Three of a modified version of 'Nanuchka I' ('Nanuchka II') with four SS-N-2B missiles have been supplied to India in 1977-78, three to Algeria in 1980-82, one to Libya in 1981, a second in February 1983, a third in February 1984 and a fourth in September 1985.

'Nanuchka I' class

Number in class: 2 + 16 (+ 3)

GERMANY (DR)
ALBIN KÖBIS
+ 3

POLAND
GORNIK
HUTNIK

Displacement, tons: 580 full load
Dimensions, feet (metres): 185.3 × 34.4 × 8.2
(56 × 10.5 × 2.5)
Missiles: SSM; 4—SS-N-2C (twin); SS-N-22 (in
one Trials 'Tarantul II'); SAM; 1—SA-N-5
(quad)
Guns: 1—76 mm (fwd); 2—30 mm Gatling (aft,
both beams)
Main engines: 2 cruising gas turbines (12 000
shp each); 2 boost gas turbines with reversible
gear box (4 000 shp each); 2 shafts: 32 000 shp
total
Speed, knots: 36 (on four turbines)
Complement: 50

The first 'Tarantul I' was completed in 1978 at
Petrovsky, Leningrad and remains in the Soviet
Navy. The single experimental 'Tarantul II' with
four SS-N-22 was completed at Petrovsky in
1981. Other 'Tarantul II' are built at Kolpino,
Leningrad and in the Pacific. A second 'Tarantul I'
without special radar is retained in USSR to train
foreign crews.

Transfers: All 'Tarantul I' class-one to Poland 28
December 1983 and second in April 1984. One to
GDR in September 1984, second in December
1984, third in September 1985 and fourth in
January 1986.

Albin Köbis *('Tarantul I' class) of the
German Democratic Republic (Fed.
German Navy, 1984)*

Number in class: 16

BAD DOBERAN
BERGEN
BUETZOW
GADEBUSCH
LÜBZ
PARCHIM
PERLEBERG
TETEROW
WAREN
WISMAR
+ 6 (+ 2)

Displacement, tons: 960 standard; 1 200 full load
Dimensions, feet (metres): 237.9 × 30.8 × 11.5 (72.5 × 9.4 × 3.5)
Missiles: SAM; 2-SA-N-5 (quad)
Guns: 2-57 mm (twin, aft); 2-30 mm twin, fwd)
A/S weapons: 2 RBU 6000; 4 single 15.8 in (400 mm) torpedo tubes; DCs
Mines: Minerails fitted
Main engines: 2 diesels; 2 shafts; 12 000 hp 25 knots
Complement: 60
Commissioned: 1981-86

Replacing 'Hai III' class. Basically very similar to Soviet 'Grisha' class. Originally classified 'Balcom 4'. Programme likely to total 18.
Other names reported: *Güstrow, Ribnitz-Darmgarten, Ludwigslust.*

Parchim ('Parchim' class) of the German Democratic Republic (Royal Danish Navy, 1982) (old number)

Number in class: 1

Displacement, tons: 280 standard; 320 full load
Dimensions, feet (metres): Hullborne: 147.6 ×
  36.1 × 9.2 *(45 × 11 × 2.8)* Foilborne: 166 × 77.1
  × 23.9 *(50.6 × 23.5 × 7.3)*
Missiles: SSM; 4—SS-N-9 (twin launchers);
  SAM; 2—SA-N-4 (twin launcher)
Gun: 1—30 mm Gatling
Main engines: 2 NK-12 gas turbines; 2 shafts; 2
  diesels; 22 000 hp = 45 knots
Complement: 35

Built at Petrovsky, Leningrad in 1975. Foil system
similar to US 'PHM', thus being the first Soviet
fully rigged hydrofoil. This is probably an R and D
craft for this system.

'Sarancha' class

Number in class: 16

Displacement, tons: 200 standard; 245 full load
Dimensions, feet (metres): 129.9 × 24.9 (41 over foils) × 6.9 (13.1 over foils) *(39.6 × 7.6 (12.5) × 2.1 (4))*
Missiles: SSM; 2—SS-N-2C (single)
Guns: 1—76 mm; 1—30 mm Gatling
Main engines: 3 M504 diesels; 3 shafts; 10 000 hp = 40 knots
Range, miles: 600 at 35 knots (foilborne); 1 500 at 14 knots (hullborne)
Complement: 30

In early 1978 the first of class was seen. Built at Izhora Yard, Leningrad. Similar hull to the 'Osa' class with similar single hydrofoil system to 'Turya' class. The combination has produced a better sea-boat than the 'Osa' class which the 'Matka' class may have been intended to replace but production appears to have stopped.

Soviet type name: Raketny kater, meaning missile cutter.

'Matka' class (Royal Danish Navy 1982)

USSR
Number in class: 60 + 35

BULGARIA
3 OSA I ; 3 (+ 1) OSA II

GERMANY (DR) **OSA I**

ALBERT GAST
FRITZ GAST
HEINRICH DORRENBACK
JOSEF SCHARES
KARL MESBERG
MAX REICHPIETSCH
OTTO TOST
PAUL EISENSCHNEIDER
PAUL WIECZOREK
RICHARD SORGE
RUDOLF EGELHOFER

POLAND **13 OSA I**

ROMANIA **6 OSA I**

Displacement, tons: 165 standard; 210 full load (245 'Osa II')
Dimensions, feet (metres): 127.9 × 25.6 × 5.9 *(39 × 7.8 × 1.8)*
Missiles: SSM; 4—SS-N-2A or B ('Osa I'); 4—SS-N-2B or C ('Osa II') (single launchers) (see *note*)
Guns: 4—30 mm/L65 (2 twin, 1 fwd, 1 aft)
Main engines: 3 diesels; M503A; 12 000 bhp (I); M504; 15 000 bhp (II) = 35 knots ('Osa I'), 37 ('Osa II')
Fuel, tons: 40
Range, miles: 400 at 34 knots ('Osa I'); 500 at 35 knots ('Osa II')
Complement: 30

'Osa I' class built in first half of the 1960s and 'Osa II' class in the latter half at a number of yards. They have a surface-to- surface missile range of up to 25 miles. Later boats have cylindrical missile launchers, comprising the 'Osa II' class which has a greater cruising range. This class was a revolution in naval shipbuilding. Although confined by their size and range to coastal operations the lethality and accuracy of the Styx missile were first proved by the sinking of the Israeli destroyer *Eilat* on 21 October 1967 by an Egyptian 'Komar' class vessel.

Soviet type name: Raketny kater, meaning missile cutter.

Transfers: 'Osa I': Algeria (3), Bulgaria (3), Cuba (5), Egypt (12—8 remaining), German Democratic Republic (15), India (8), Iraq (4), North Korea (8), Poland (14), Romania (6), Syria (6), Yugoslavia (10).
'Osa II': Algeria (9), Angola (6), Bulgaria (4), Cuba (13), Ethiopia (4), Finland (4), India (8), Iraq (8), Libya (12), Somalia (2), Syria (10—six remaining), North Yemen (2) (subsequently returned), South Yemen (8), Viet-Nam (8).

Left: *'Osa II' class*
Overleaf
Left: *'Osa I' class of the German Democratic Republic, 1980*
Right: *'Osa I' class (MoD, 1980)*

Number in class: 30

Displacement, tons: 190 standard; 250 full load
Dimensions, feet (metres): 129.9 × 24.9 (41 over foils) × 5.9 (13.1 over foils) *(39.6 × 7.6 (12.5) × 1.8 (4))*
Guns: 2—57 mm/L70 (twin, aft); 2—25 mm/L80 (twin, fwd)
Torpedo tubes: 4—21 in *(533 mm)*
A/S weapons: DCs
Main engines: 3 M504 diesels; 3 shafts; 15 000 shp
Speed, knots: 40 (foilborne)
Range, miles: 600 at 35 knots (foilborne); 1 450 at 14 knots (hullborne)
Complement: 30

The second class of hydrofoil with single foil forward (after 'Matka' class). Has a naval orientation rather than the earlier 'Pchela' class of the KGB. Entered service from 1971—built at Leningrad and Vladivostok. Now produced only for export. Basically 'Osa' hull.

Soviet type name: Torpedny kater, meaning torpedo cutter.

Transfers: Two to Cuba 9 February 1979, two in February 1980, two in February 1981, two in January 1983 and one in November 1983; one to Ethiopia in early 1985; one to Kampuchea in mid 1984 and one in early 1985; two to Viet-Nam in mid-1984, one in late 1984 and two in early 1986.

'Turya' class (Federal German Navy, 1982)

Number in class: 22

VT 51-72

Displacement, tons: 39 standard; 45 full load
Dimensions, feet (metres): 71.5 × 20.7 oa; 11.8
  hull × 3.3 *(21.8 × 6.3 oa; 3.6 hull × 1)*
Guns: 4—14.5 mm (twin)
Torpedo tubes: 2—21 in *(533 mm)*
Main engines: 3 M50 diesels; 3 600 hp = 50
  knots (foilborne in calm conditions)
Range, miles: 500 cruising
Complement: 11

Hydrofoils of the same class as the Chinese
which were started in 1956.
  Three imported from China. Remainder locally
built in a programme of about two a year which
started 1973-74 and is now complete. Some
have had guns and foils removed.

*'Huchuan' class of the Romanian
Navy*

USSR

Number in class: 10

BULGARIA **6**

GERMANY (DR) 15
ADAM KUCKHOFF
ARTHUR BECKER
ARVID HARNACK
BERNARD BÄSTLEIN
E. ANDRÉ
ERICH KUTTNER
ERNST GRUBER
ERNST SCHNELLER
FRITZ BEHN
FRITZ HECKERT
HEINZ KAPELLE
JOSEF ROEMER
MAX ROSCHER
WILHELM FLORIN
WILLI BANSCH

Displacement, tons: 145 standard; 170 full load
Dimensions, feet (metres): 113.8 × 22 × 4.9
   (34.7 × 6.7 × 1.5)
Guns: 4—30 mm/L65 (twin)
Torpedo tubes: 4—21 in *(533 mm)* (single)
A/S weapons: 2 DC racks (12 DCs)
Mines: Capacity for 6
Main engines: 3 M503A diesels; 3 shafts; 12 000
   bhp = 45 knots
Fuel, tons: 30
Range, miles: 850 at 30 knots; 460 at 42 knots
Complement: 23

First of class produced in 1962. Programme
completed by 1974.
The 'Mol' class was an export version of the
'Shershen'.

Soviet type name: Torpedny kater, meaning
torpedo cutter.

Transfers: Angola (4), Bulgaria (6), Cape Verde
(2), Congo (3), Egypt (6), German Democratic
Republic (18), Guinea (2), Guinea-Bissau (2),
North Korea (3), Viet-Nam (18), Yugoslavia (2 + 13
locally built).
   'Mol' class: Ethiopia (2 in 1978), Somalia (4 in
1976-77), Sri Lanka (1 in 1975).

Note: Two 'Mols' in Somalia and one in Sri Lanka
do not have torpedo tubes.

'Shershen' class (1974)

Number in class: 31

KARL BAIER
+ 30

Displacement, tons: 30 full load
Dimensions, feet (metres): 59 × 16.4 × 6.6 *(18 × 5 × 2)*
Guns: 2—23 mm (twin)
Torpedo tubes: 2—21 in *(533 mm)* (stern launching)
Main engines: 3 M50-F4 diesels; 3 shafts; 3 600 hp = 40 knots

A class first reported in 1975. Can be used for minelaying and commando operations. Conversion for new tasks is a speedy job. One trials boat, V 87 (unarmed). Production completed 1981-82.

'Libelle' class of the German Democratic Republic (1978)

Number in class: 7

Displacement, tons: 70 full load
Dimensions, feet (metres): 82 × 18 × 6 *(25 × 5.5 × 1.8)*
Guns: 2—30 mm (twin)
Torpedo tubes: 4—21 in (533 mm)
Main engines: 4 diesels; 4 shafts; 4 800 hp = 34 knots

Polish built since early 1970s.

'Wisla' class of the Polish Navy

Number in class: 6

Displacement, tons: 22 standard; 25 full load
Dimensions, feet (metres): 62.3 × 10.8 × 3.3 *(19 × 3.3 × 1)*
Guns: 2—14.5 mm (twin)
Torpedo tubes: 2—18 in *(457 mm)*
A/S weapons: 4—8 DCs
Main engines: 2 M50 diesels; 2 shafts; 2 400 bhp = 40-42 knots
Fuel, tons: 3
Range, miles: 410 at 30 knots
Complement: 12

Built in the Soviet Union in 1955-56. Obsolete and ready for deletion. One has had guns removed.

'P-4' class of the Romanian Navy

Number in class: 1

Displacement, tons: 400 full load
Dimensions, feet (metres): 164 × 27.9 (33.5 over
   foils) × 13.1 (19.4 foils) *(50 × 8.5 (10.2) × 4 (5.9))*
Guns: 2—30 mm Gatling
A/S weapons: 8—16 in *(406 mm)* A/S torpedo
   tubes (quad, fwd)
Main engines: CODOG; 3 NK-12 gas turbines;
   30 000 hp; 2 diesels = 45 knots
Complement: 45

This is a much larger hydrofoil than any
previously seen. First sighted 1977. Unusual
features include a new hydrofoil arrangement
with a single fixed foil forward, the large gas
turbine exhausts aft, the trainable torpedo
mountings forward and the large object forward
of the first Gatling. Probably used for research
and development.

'Babochka' class (1978)

Number in class: 8

Displacement, tons: 70 standard; 80 full load
Dimensions, feet (metres): 83 × 19 × 4.3 *(25.3 × 5.8 × 1.3)* (without foils)
Guns: 4—23 mm/L87 (twin)
A/S weapons: 6 DCs
Main engines: 2 diesels; 2 shafts; 6 000 bhp = 45-50 knots
Complement: 12

This class of hydrofoil was built in the mid-1960s. Used for frontier guard duties by KGB in Baltic and Black Seas. Numbers reducing due to age.

Left: *'Pchela' class (Royal Danish Navy, 1984)*
Right: *'Pchela' class (US Navy, 1977)*

Number in class: 19

VP TYPE

VS TYPE

SATURN
VENUS

Displacement, tons: 120 standard; 155 full load
Dimensions, feet (metres): 128 × 18 × 5.6 *(39 × 5.5 × 1.7)*
Guns: VP type: 1—57 mm; 2—37 mm/L63 (twin). VS type: 1—37 mm; 4—14.5 mm MGs
A/S weapons: VS type: two 5-barrelled RBU 1200; 2 DC racks
Main engines: 4 diesels; 4 220 bhp = 30 knots
Range, miles: 800 at 17 knots
Complement: 35

Three variants of the 'Shanghai' class of which the VS type (patrol A/S) was a new departure and the two named craft differ greatly in their bridge superstructure. Built at Mangalia since 1973 in a programme of about two a year (which is now complete) with the exception of VP 22, 24 and 25 which were imported from China.

'Shanghai' class of the Romanian Navy

Number in class: 18 + ?

Displacement, tons: 580 full load
Dimensions, feet (metres): 190.3 × 34.4 × 8.2
 (58 × 10.5 × 2.5)
Missiles: SAM; SA-N-5 (quad) (8 missiles)
Guns: 1—76 mm/L59; 1—30 mm Gatling
Torpedo tubes: 4—16 in *(406 mm)* (single)
A/S weapons: 2 RBU 1200; 2 DC racks
Main engines: 4 diesels; 16 000 hp = 28-34 knots
Complement: 80 (?)

First laid down in 1977 and completed in 1979. This appears to be an A/S version of the 'Tarantul' class having the same hull form with a two metre extension for dipping sonar. Replacement for 'Poti' class. In series production.

First three of class have a lower bridge than successors.

Soviet type name: Maly protivolodochny korabl, meaning small anti-submarine ship.

Left: *'Pauk' class (1981)*
Right: *'Pauk' class (1981)*

Number in class: 1

Displacement, tons: 210 full load
Dimensions, feet (metres): 127.9 × 26.6 × 5.9
(39 × 8.1 × 1.8)
Guns: 1—76 mm; 1—30 mm Gatling
Main engines: 3 M503A diesels; 3 shafts; 12 000
shp = 36 knots
Complement: 30

Built on an 'Osa' hull, this single craft has the
tripod mast mounted further aft and different
radar. Built at Petrovsky, Leningrad in 1970 with
twin 57 mm, 30 mm Gatling and prototype Bass
Tilt as trials ship for 'Grisha III' class. In 1975 the
57 mm replaced by single 76 mm turret as trials
ship for 'Matka' and 'Nanuchka III' classes. Also
carried out trials for chaff launchers for 'Krivak'
and 'Nanuchka' classes.

'Slepen' class (with 57 mm guns)

Number in class: 59

Displacement, tons: 400 full load
Dimensions, feet (metres): 196.8 × 26.2 × 6.6
 (60 × 8 × 2)
Guns: 2—57 mm (twin mounting)
Torpedo tubes: 4—16 in *(406 mm)*
A/S weapons: 2 RBU 6000
Main engines: 2 gas turbines; 30 000 shp; 2
 M503A diesels; 8 000 shp; 2 shafts = 38 knots
Range, miles: 4 500 at 10 knots; 500 at 37 knots
Complement: 80

This class of ship was under series construction from 1961 to 1968 at Zelenodolsk and Khabarov and is now being replaced by the 'Pauk' class.

Soviet type name: Maly protivolodochny korabl meaning small anti-submarine ship.

Transfers: Three to Bulgaria (mid-1970s), three to Romania (late 1960s).

Left: *'Poti' class (1979)*
Right: *'Poti' class (1981)*

Number in class: 100

Displacement, tons: 170 standard; 210 full load
Dimensions, feet (metres): 127.9 × 25.6 × 5.9
  (39 × 7.8 × 1.8)
Guns: 4—30 mm/L65 (twin)
Torpedo tubes: 4—16 in *(406 mm)*
A/S weapons: 2 DC racks
Main engines: 3 M503A diesels; 12 000 bhp = 36
  knots
Range, miles: 800 at 24 knots; 500 at 35 knots
Complement: 30

Based on the hull design of the 'Osa' class.
Construction started in 1967 and is still
continuing at Petrovsky, Leningrad and
Vladivostok for the KGB (which operates the
whole class) at a rate of about five a year.

Soviet type name: Pogranichny storozhevoy
korabl, meaning border patrol ship.

Left: *'Stenka' class (L & L van
Ginderen, 1983)*
Right: *'Stenka' class (Federal
Germany Navy, 1982)*

Number in class: 15

Displacement, tons: 170 standard; 215 full load
Dimensions, feet (metres): 137.8 × 19.7 × 5.9
(42 × 6 × 1.8)
Guns: 4—25 mm (2 twin mountings) (see notes)
A/S weapons: 4 RBU 1200; 2 DC racks
Torpedo tubes: 2—16 in *(406 mm)* (some)
Mines: Can carry 18
Main engines: 3 diesels; 3 shafts; 7 500 bhp = 28
knots
Range, miles: 1 100 at 13 knots; 350 at 28 knots
Complement: 31

Built between 1957 and late 1960s at
Zelenodolsk and Khabarov— total about 150.
Steel hulled. Some of this class have one 57 mm
and two 25 mm guns. A few have been
modernised with two 16-in anti-submarine
torpedo tubes and only two 25 mm guns. This
class is being phased out of service, although
some are operated by the KGB.

Soviet type name: Maly protivolodochny korabl,
meaning small anti-submarine ship.

Transfers: Algeria (6), Bulgaria (6 in 1963), Cuba
(10), Egypt (12), German Democratic Republic
(12—all now scrapped), Iraq (3), North Korea (15),
Mozambique (2 in mid 1985), Viet-Nam (13 of
which 5 have been lost), South Yemen (2).

'SO-1' class (1981)

Number in class: 16

KIROV
KORUND
PAVEL VINOGRADOV
PRIMORSKY KOMSOMOLETS
ST. LEITENANT VLADIMIROV
+ 11

Displacement, tons: 790 standard; 860 full load
Dimensions, feet (metres): 229.9 × 29.5 × 7.9
    (70.1 × 9 × 2.4)
Guns: 4—57 mm (twin); 4—25 mm (twin) (in some)
A/S weapons: 2 RBU 1200; 2 DCTs
Mines: 18
Main engines: 2 diesels; 2 shafts; 4 000 bhp $_5$ 17 knots
Range, miles: 2 500 at 13 knots
Complement: 82

Built from 1957 to 1963, conversion to patrol ships in about 1975. Originally fleet minesweepers with steel hulls. Of this class 12 were converted to submarine rescue ships with armament and sweeping gear removed. Frequently deployed to the Indian Ocean. Have minelaying capability. A number is employed by the KGB, these being named.

Soviet type name: Storozhevoy korabl, meaning patrol ship or, when used by KGB, pogranichny storozhevoy korabl, meaning border patrol ship.

'T-58' class (MoD, 1981)

Number in class: 8

NARVIK
WYTRWALY
ZAWZIETY
ZRECNY
+ 4

Displacement, tons: 150
Dimensions, feet (metres): 137.8 × 19 × 6.6 *(42 × 5.8 × 2)*
Guns: 4—30 mm
A / S weapons: 2 internal DC racks
Main engines: 2 diesels = 20 knots

Slightly smaller than original 'Obluze' class. Built late 1960s.

Note: All Polish-built Large Patrol Craft based on German R-boat design with modified superstructure and deckhouses.

'Mod' Obluze' class of the Polish Navy (1970)

Number in class: 3

Displacement, tons: 310 standard; 380 full load
Dimensions, feet (metres): 170.9 × 21.3 × 6.9
  (52.1 × 6.5 × 2.1)
Guns: 1—85 mm; 2—37 mm (single); 6—12.7
  mm (twin)
A/S weapons: 2 DC throwers; 2 DC racks
Main engines: 3 diesels; 3 shafts; 3 300 bhp = 24
  knots
Range, miles: 1 500 at 12 knots
Complement: 50

Transferred by the USSR in 1956. Survivors of a very numerous class built in Soviet yards in the early 1950s.

'Kronshtadt' class of the Romanian Navy

Number in class: 2 + 1

IVAN ROGOV
ALEKSANDR NIKOLAYEV
+ 1

Displacement, tons: 13 000 full load
Dimensions, feet (metres): 521.6 × 80.2 × 21.2
(27.8 flooded) *(159 × 24.5 × 6.5 (8.5))*
Aircraft: 4 Helix helicopters
Missiles: SAM; 2—SA-N-4 (twin launcher) (20
missiles); 2—SA-N-5 (quad) *(A. Nikolayev)*
Guns: 2—76 mm (twin); 4—30 mm Gatlings; 1—
122 mm rocket launcher BM-21 (naval) (2 ×
20-barrelled)
Main engines: 2 gas turbines; 2 shafts;
40 000 shp = 25 knots
Range, miles: 4 000 at 18 knots
Complement: 400

First appeared in 1978, having been built at
Kaliningrad. Second completed in 1983. Has bow
ramp with beaching capability leading from a
tank deck 200 ft long and 45 ft wide. Stern doors
open into a docking bay 250 ft long and 45 ft
wide. A helicopter spot forward has a flying-
control station and the after helicopter deck and
hangar is similarly fitted. With a capacity for a
battalion of naval infantry (522) and up to 20
tanks as well as supporting vehicles, this class
provides a long range, long endurance assault
capacity of far greater potential than any previous
Soviet ship. Can carry two 'Lebed' ACVs and one
'Ondatra' class LCM in docking bay. Third ship
laid down 1985.

Soviet type name: Bolshoy desantny korabl,
meaning large landing ship.

Aleksandr Nikolayev *('Ivan Rogov'
class) (JMSDF 1985)*

INDEXES